AF618761

„Fortschritte der Hochpolymeren-Forschung/Advances in Polymer Science"

erscheinen zwanglos in einzeln berechneten Heften, die zu Bänden vereinigt werden.

Sie enthalten Fortschrittsberichte monographischen Charakters aus dem Gebiet der Physik und Chemie der Hochpolymeren mit ausführlichen Literaturzusammenstellungen. Sie sollen der Unterrichtung der auf diesen Gebieten Tätigen über solche Themen dienen, die in letzter Zeit besondere Aktualität gewonnen haben, bzw. die in neuerer Zeit eine lebhafte und nach literarischer Zusammenfassung verlangende Entwicklung erfahren haben.

Anschriften der Herausgeber:

Prof. Dr. J. D. Ferry, Department of Chemistry, The University of Wisconsin, Madison, 6 Wisconsin/USA.

Prof. Dr. C. G. Overberger, Polytechnic Institute of Brooklyn, 99 Livingston Street, Brooklyn 1, New York/USA.

Prof. Dr. G. V. Schulz, Institut für physikalische Chemie der Universität, Mainz.

Prof. Dr. A. J. Staverman, Fruinlaan 6, Leiden/Holland.

Prof. Dr. H. A. Stuart, Institut für physikalische Chemie der Universität, Mainz.

Springer-Verlag

Heidelberg	**Berlin W 35**
Neuenheimer Landstraße 28—30	Reichpietschufer 20
Fernsprecher 27901	Fernsprecher 130131

1. Band **Inhaltsverzeichnis** 1. Heft

ISBN 978-3-540-02242-8 ISBN 978-3-540-36888-5 (eBook)
DOI 10.1007/978-3-540-36888-5

Fortschr. Hochpolym.-Forsch., Bd. 1. S. 1—34 (1958)

Specific Ion Binding by Polyelectrolytes

By

HERBERT MORAWETZ

Department of Chemistry, Polytechnic Institute of Brooklyn[1]

With 1 Figure

Table of Contents

a) Introduction

Polyelectrolytes are macromolecules carrying large numbers of ionizable groups. Some essential components of living organisms, particularly the proteins and nucleic acids, fall into this category. DOTY and EHRLICH have reviewed the intensive work carried out in recent years on synthetic polyelectrolytes, which was stimulated in part by the hope that they would serve as models furthering our understanding of the naturally occurring materials. While a protein or a nucleic acid contains a number of different types of chain segments arranged in a sequence which is generally unknown, synthetic polyelectrolytes of great chemical simplicity, containing a single repeating unit, may easily be prepared. On the other hand, the specific folding of the polypeptide chains in globular proteins and the helical structure of nucleic acid demonstrated by WATSON and CRICK produce macromolecules whose shape in solution is much more closely defined than that of the highly flexible synthetic polymers.

When the polyion carries a high net electrical charge, the electrostatic interaction between polyion and counter-ions may result in unusually small ionic activity coefficients. This effect has been treated theoretically by KATCHALSKY and LIFSON, by OSAWA, IMAI and KAGAWA and by MARCUS, and studied experimentally by KERN, by KAGAWA and KATSUURA (1952, 1955) and by KATCHALSKY and LIFSON. It will disappear for polyampholytes at the isoelectric point and may be eliminated

[1] Prepared during the tenure of a LOUIS LIPSKY Exchange Fellowship at the Weizmann Institute of Science, Rehovoth, Israel.

in solutions of polymeric acids or bases by addition of simple electrolytes. It is, however, important to realize that highly specific interactions between polyion and counter-ions may be superimposed on electrostatic effects. Groups participating in complex ion formation, referred to as ligand groups, may include negatively charged anionic sites such as carboxylate, uncharged basic nitrogen groups, or groups such as aliphatic hydroxyl, which do not participate in ionization equilibria.

Before considering the interpretation of the specific ion binding by molecules carrying large numbers of ligand groups it is essential to survey the known principles of complex ion formation with small molecules and methods available for their investigation. Since the excellent monograph by MARTELL and CALVIN was published, notable advances have been made, particularly in the thermodynamics of complex formation and the interpretation of complex ion spectra. The understanding of these well defined complexes will be, naturally, much more exact than what can be hoped for with polymers. The problem will be particularly difficult when several groups of a polymer participate in the binding of a single ion, since the probability and the energetics of such cooperative effects necessitate a detailed understanding of macromolecular configuration. The high specificity of enzyme requirements for activating ions cannot be visualized without such a critical juxtaposition of a number of ligand groups and it is intriguing to speculate to what extent a similar degree of specificity can be built into a synthetic polyelectrolyte.

The ion binding of proteins has been reviewed by KLOTZ, by SCATCHARD et al. (1954) and by GURD and work on the ion activation and inhibition of enzymes has been summarized by LEHNINGER, but important new evidence has been reported since these reviews were published. The study of ion binding of synthetic polyelectrolytes is, by contrast, a recent development. It is hoped that a unified treatment of problems posed by complex ion formation with small molecules and with both natural and synthetic polymers will point to previously neglected methods of investigation and stimulate new research in this field.

b) Complex Ion Formation with Simple Ligands

WERNER's classical work on complex salts led to the preparation of a large number of these substances in the crystalline state and established some important principles of their formation. It was found that complex forming cations are characterized by a "coordination number" specifying the number of ligand groups such as NH_3, CN^-, I^-, CH_3COO^-, H_2O etc., which combine with the cation to form distinct chemical species. This "coordination number" was 6 in most cases, but it was 2 for Ag^+ and in some cases it was 4 (e. g. Cu^{++}, Cd^{++}, Zn^{++}) or 8 (Mo^{+4}, W^{+4}). Moreover, a study of the isomerization of complex salts as well as X-ray crystal analysis proved the existence of a definite geometrical pattern in the distribution of the ligand groups around the central ion. Thus, silver ion was shown to form a 180° bond angle with its two ligands. Four ligand groups could be held either in a tetrahedral (e. g. $Cd^{++}Zn^{++}$) or a square planar structure (e. g. Cu^{++}, Pt^{++}) and six ligands formed a regular octahedron around the central ion (MARTELL and CALVIN).

In recent years, the emphasis in the study of complex ions has shifted from the crystalline state to their behavior in solution, particularly since J. BJERRUM (1941) developed techniques for the study of the association equilibria of cations with successive ligand molecules (MARTELL and CALVIN). Although the association of a cation M^{+m} with a ligand A^{+n} involves the displacement of a coordinated water molecule

$$[M(H_2O)_N]^{+m} + A^{+n} \leftrightarrows [M(H_2O)_{N-1}A]^{m+n} + H_2O \qquad (1)$$

we shall in the following omit in our notation both the water molecules involved in the equilibria and the charges of the various species. The equilibria are then defined by "formation constants" of the form

$$\frac{(MA)}{(M)\ (A)} = K_1\,; \quad \frac{(MA_2)}{(MA)\ (A)} = K_2\,; \quad \cdots \frac{(MA_i)}{(MA_{i-1})\ (A)} = K_i\,. \qquad (2)$$

Every cation has a "characteristic coordination number" N, corresponding to the number of ligand groups bound in a similar manner; additional groups may sometimes be bound much more weakly [BJERRUM (1950)]. Thus, for Cu^{++} the characteristic coordination number is 4, although $[Cu(NH_3)_5]^{++}$ will be present, to some extent, in very concentrated ammonia solutions [BJERRUM (1950); BJERRUM, BALLHAUSEN and JØRGENSEN].

Four main methods are used most frequently in the determination of complex ion formation constants:

(1) BJERRUM's titration method depends on the competition of the cation M and hydrogen ion for the ligand A, defining the equilibria

$$(HA) + (MA_{i-1}) \leftrightarrows (H) + (MA_i)\,.$$

The titration curve of HA will then be displaced to lower p_H in the presence of the complex forming cation M. Writing C_A for the stoichiometric ligand concentration, α for the degree of neutralization of its conjugate acid and $(X) = \sum_i i\,(MA_i)$ for the concentration of the ligand in the various complex ions formed,

$$(A) = C_A\,\alpha + (H) - (OH) - (X)$$
$$C_A = (HA) + (A) + (X) \quad (3)$$
$$(H)\,(A)/(HA) = K_a$$

allowing the calculation of (X). The average number of ligands complexed with a cation present in a stoichiometric concentration C_M is then $\bar{n} = (X)/C_M$. Provided the successive formation constants are far apart, $(A) = 1/K_1$ for $\bar{n} = 0.5$, $(A) = 1/K_2$ for $\bar{n} = 1.5$, etc. The treatment for the general case, when the successive formation constants may be close to one another, is given by BJERRUM (1941) and by MARTELL and CALVIN.

It should be noted that use of the titration method leads to "concentration constants" defined for a given ionic strength rather than thermodynamic equilibrium constants. The latter may be obtained by an extrapolation procedure or an estimate of activity coefficients which is, however, hampered by uncertainty about a suitable value for the distance of closest approach to the complex ion (IZATT et al.). It is also clear that use of relations (3) presupposes that no p_H shift takes place due to a change of ionic activity coefficients when the complex forming ion is introduced. This assumption is well justified when the formation constants are large and pronounced p_H shifts may be obtained on addition of small amounts of complex forming cations, but in the study of very weak complexes such as the aceto-alkaline earth (CANNAN and KIBRICK; NANCOLLAS) or aceto-silver ion (MACDOUGALL and PETERSON) great care must be taken to differentiate between effects due to changing activity coefficients and those resulting from complex formation.

Table 1. *Equilibrium Constants for Cation Hydrolysis*

M	$[M(H_2O)_n]^{+m} \rightleftarrows [M(H_2O)_{n-1}(OH)]^{+m-1} + H^+$		
	$-\log K$	M	$-\log K$
Ba^{++}	13.4	Fe^{++}	9.5
Sr^{++}	13.0	Pb^{++}	8.8
Ca^{++}	12.7	Cd^{++}	8.5
Ag^{+}	11.7	Cu^{++}	8.0
Mg^{++}	11.4	Al^{+++}	4.9
Ni^{++}	10.6	Cr^{+++}	3.9
Co^{++}	10.2	Hg^{++}	3.7
Zn^{++}	9.6	Fe^{+++}	3.0

The titration method cannot be employed if the complex ion is stable at relatively high acidity, since the determination of the hydrogen ions displaced from the conjugate acid of the ligand by the complexing cation cannot be carried out with sufficient precision. Conversely, the titration method gives results which are difficult to interpret whenever a given complex forms only at a relatively high p_H, where the cation also tends to associate with hydroxyl ions. Table 1 reproduces MATTOCK's compilation of the first hydrolysis constants of some cations whose complexes are commonly studied. It should be noted that the constants obtained by different authors often show rather poor agreement and that in addition to the species [MOH] the mixed complex containing both hydroxyl and other ligands may have to be considered.

(2) Potentiometric measurements employing an electrode of the metal whose complexes are being studied provide a convenient method for the direct determination of the free cation activity (MacDougall and Peterson). The method is, however, limited by difficulties in obtaining reversible electrodes of some of the metals which are most interesting in their complexation behavior [Bjerrum (1941)].

(3) When complex ion formation results in an absorption change in an easily accessible range of the spectrum, colorimetric measurements may provide the most convenient tool for following the reaction. This will be the case particularly when a single equilibrium has to be considered, such as $Fe^{+++} + OH^- \rightarrow [Fe\,OH]^{++}$ (Siddall and Vosburgh), or $[Cu\,en_2]^{++} + OH^- \rightarrow [Cu\,en_2OH]^+$ [Jonassen, Reeves and Segal (1955b)], although Woldbye has shown that the method is applicable, in principle, also for determination of successive formation constants.

(4) The activity of the free cation may be fixed by employing a second phase acting as a "cation buffer". This buffer may be a sparingly soluble salt (Yonte and Martin; King) but the work of Fronaeus (1951) and of Schubert has shown that ion exchangers are particularly convenient for this purpose. If a radioactive isotope of the cation is used, it may be employed at extremely low concentration so that the amount of ligand present in the various complexes is negligible compared to the total ligand in solution. If the stoichiometric concentration of M in solution is C_M and C_M° in the presence and the absence of the ligand, respectively, then

$$C_M - C_M^\circ = \Sigma(\mathrm{MA_i}) \tag{4}$$

and since $C_M^\circ = (\mathrm{M})$ we may use the relations (2) to obtain

$$\Sigma(\mathrm{MA_i}) = (\mathrm{M})\ [K_1(A) + K_1K_2(A)^2 + \cdots]\,. \tag{5}$$

Setting $(\mathrm{A}) = C_A$,

$$\frac{C_M - C_M^\circ}{C_M^\circ} = K_1C_A + K_1K_2C_A^2 + \cdots \tag{6}$$

yielding the successive formation constants from determination of the cation concentration as a function of C_A at constant ionic strength.

If we compare the successive association equilibria of e. g. Cu^{++} and NH_3, it is clear, that the first ammonia molecule may add to any one of four coordinate positions, while NH_3 may dissociate from $CuNH_3^{++}$ in only one way. The second complex formation equilibrium involves the possible addition of NH_3 to the three free coordinate positions of $CuNH_3^{++}$ and two possible ways in which NH_3 may dissociate from $[Cu(NH_3)_2]^{++}$. Thus, we should expect, on statistical grounds alone, a ratio of 4/(3/2) for K_1/K_2. In general, we should obtain in the absence of energetic interactions between the ligand groups, for the i-th formation constant

$$K_i = [(N - i + 1)/i]\, K_0 \tag{7}$$

where K_0 represents the hypothetical ligand association constant with a given coordination site of the cation. BJERRUM's (1950) data reproduced in Table 2 show clearly that the statistical correction does not suffice to

Table 2.

Intrinsic Association Constants for Successive Stages of Metal-Ammine Formation

Cation	($\log K_0$)						
	N	$i = 1$	$i = 2$	$i = 3$	$i = 4$	$i = 5$	$i = 6$
Ag^+	2	2.90	4.13				
Zn^{++}	4	1.77	2.26	2.60	2.75		
Cu^{++}	4	3.55	3.32	3.07	2.73		
Ni^{++}	6	2.02	1.84	1.60	1.32	1.15	0.81

explain the trend in the successive metal ammine formation constants and that, in fact, the presence of an associated ammonia molecule may favor further ammonia addition, as with Ag^+ and Zn^{++}, or hinder it, as with Cu^{++} and Ni^{++}. It is also significant that in spite of the repulsion between NH_3 groups in the copper ammines as inferred from the trend of the K_0 values, BJERRUM, BALLHAUSEN and JØRGENSEN assigned, on spectroscopic evidence, to the ammonia groups in $[Cu(NH_3)_2(H_2O)_2]^{++}$ the *cis*, rather than the *trans* position in the square planar complex. IRVING and WILLIAMS (1953) point out that the assumption of a statistical relation between successive complex formation constants assumes that (1) the size of the ligand is similar to that of the displaced water molecule, (2) dipole interaction between two ligand molecules is the same as between ligand and water and (3) there is no electronic rearrangement at any stage of complex formation. It is thus hardly surprising that most cases seem to deviate from "ideal" behavior.

When the ligand groups are negatively charged, their initial repulsion would lead one to expect a downward trend in the K_0 values of successive formation constants. This expectation is realized in most cases, although some striking specific effects may occasionally lead to anomalous results. Thus, the $\log K_0$ values for the formation of the successive iodo complexes of Cd^{++} according to LEDEN (1941) are 1.48, 0.59, 2.33, 2.08. In view of the many factors involved, it is doubtful how much significance may be properly attached to calculation of effective ionic radii from the ratio of successive formation constants [N. BJERRUM (1926)] in cases which are apparently "well behaved".

The nature of the bond between the central ion and the coordinated groups has been dealt with in two different ways. PAULING interpreted the behavior of transition metal complexes by postulating covalent bonds formed by electron pairs donated by the ligands to empty orbitals of the central ion. This concept was supported by changes in the magnetic properties of some ions on complex formation and such changes were used, in fact, to differentiate between ionic and covalent complexes. Chemical evidence included the dramatic change in the oxidation potential of Co^{++}/Co^{+++} on complexation with 6 cyanide groups; whereas Co^{+++} is one of the most powerful oxidizing agents, $[Co(CN)_6]^{---}$ will reduce water with evolution of hydrogen. This change is interpreted as due to the fact that on accepting 12 electrons of 6 CN^- groups, one electron in $[Co(CN)_6]^{----}$ has to be promoted to the relatively high energy 5s orbital.

In recent years, another theory has been found particularly fruitful in the interpretation of transition metal complex ion spectra. It was observed that relatively low intensity absorption bands ($\log\varepsilon \approx 0$—2) in the visible spectrum are shifted on complex formation to shorter wavelength, while intense bands in the ultraviolet ($\log\varepsilon \approx 4$) move in the opposite direction (HARTMANN and SCHLÄFER). Some discussion of the theory of these spectral shifts seems to be in order, since it will be shown later that similar effects were observed with Cu^{++} in the presence of polymeric acids. ILSE and HARTMANN pointed out that the two types of absorption bands are characteristic of the central ion and the ligand, respectively, suggesting that each retains much of its individuality in the complex. The visible bands, due to electronic transitions between the 3d orbitals of the central ion, were interpreted by a theory due to BETHE, showing how the 3d energy levels are split by an applied crystal field in a manner depending on the field symmetry. The method was first applied by ILSE and HARTMANN to Ti^{+++} having a single 3d electron and was later used with highly satisfactory results to interpret spectra of complex ions of other transition metals (BJERRUM, BALLHAUSEN and JØRGENSEN; BALLHAUSEN and JØRGENSEN; JØRGENSEN). The ultraviolet bands, characteristic of the ligand, were studied extensively by LINHART and WEIGEL (1951a, b; 1955) and were found to be shifted to longer wavelength by a distance characteristic of the central ion, but independent of the ligand used. The evidence pointed to these bands originating from electronic transitions from the ligand to the central ion. In spite of these apparent differences, ORGEL has claimed that PAULING's approach and the crystal field theory of complex ions lead essentially to equivalent results with a strong crystal field corresponding to "covalent bonding". It was also shown (HARTMANN and SCHLÄFER; BALLHAUSEN and JØRGENSEN) that the crystal field theory may account for changes in the magnetic properties of the cation attending the formation of some complex ions.

Since a ligand functions as a base in combining with a proton or any other cation, a correlation between the basicity of a ligand and the stability of its complex ions would be expected. Such correlations are, in fact, found for series of similar ligands. MILBURN found that with a number of *p*- and *m*-substituted phenols, the equilibrium constant for

$$XC_6H_4OH + Fe^{+++} \rightleftarrows [XC_6H_4OFe]^{++} + H^+$$

is 0.11, independent of the substituent. On the other hand, silver ion association constants of amines k_{NAg} vary within a much narrower range than their hydrogen ion association constants k_{NH}. Nevertheless, BJERRUM (1950) made a very useful generalization in showing that $\log k_{NAg}/\log k_{NH}$ lies close to 0.35 for amines whose $\log k_{NH}$ covers all the range from 2 to 11.

A recent survey by WILLIAMS of thermodynamic data on complex ion formation has shown that the entropy change favors the complex if its formation involves charge neutralization. This is explained by the high degree of order of water molecules in the neighborhood of ions. Entropy effects will be decisive for the formation of complexes from anions of strong acids, which would not be expected to exhibit a strong energetic interaction with cations. Another interesting aspect of the entropy of complex formation was discussed by FRANK and EVANS, who pointed out that the displacement of water by ammonia to form $[Ni(NH_3)_6]^{++}$ is characterized by $\Delta S = -22$ e. u. while the corresponding reaction with methylamine to form $[Ni(NH_2CH_3)_6]^{++}$ has $\Delta S = +73$ e. u. This striking discrepancy was interpreted by assuming that non-polar residues, such as the methyl group of methylamine, stabilize an ice-like structure of water which melts when the "iceberg" is brought to the vicinity of an ionic charge. As for the ΔH of complex formation, WILLIAMS proposed that it may be represented by the sum of three terms accounting for (1) the electrostatic interaction proportional to Z/r, the ratio of the ionic charge and radius (2) the tendency of the cation to accept electrons from the ligand, proportional to the cation ionization potential and (3) the repulsive potential between ligand groups proportional to $1/r^3$. The first term is decisive for the alkaline earths, while the second one seems to govern the tendency of transition metal cations to engage in complex formation. The comparison of the first complex formation constants of a variety of cations with acetate and ammonia is given in Table 3 and shows that the stability of the ammonia complexes is far more sensitive to the nature of the cation. With the alkaline earths, acetate forms more stable complexes while ammonia, the better electron donor, is the stronger complexing agent with the transition metal cations.

The rate of ligand exchange varies within wide limits from reactions which are too fast to be followed by conventional techniques to others which are extremely slow. TAUBE, in an extensive review of this field, has contrasted the behavior of $[Cu(NH_3)_4]^{++}$, which decomposes instantaneously in acid solution with that of $[Co(NH_3)_6]^{+++}$, which persists almost indefinitely in 1 M HCl. Similarly $[Fe(CN)_6]^{----}$ is stable for extended periods in 1 M acid, although only 10^{-12} of Fe^{++} is complexed at equilibrium. There is no general correlation between the thermodynamic stability of complex ions and the rates of their reactions.

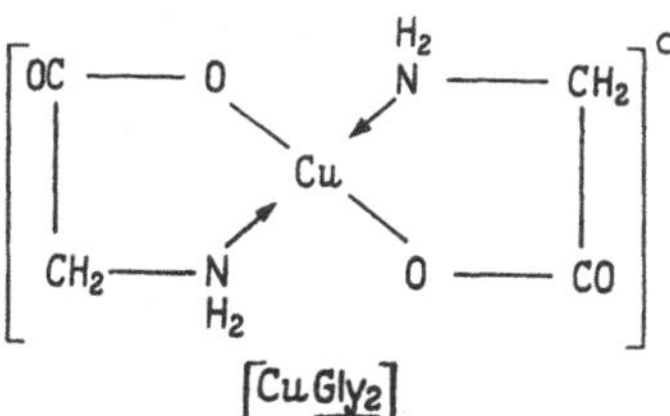

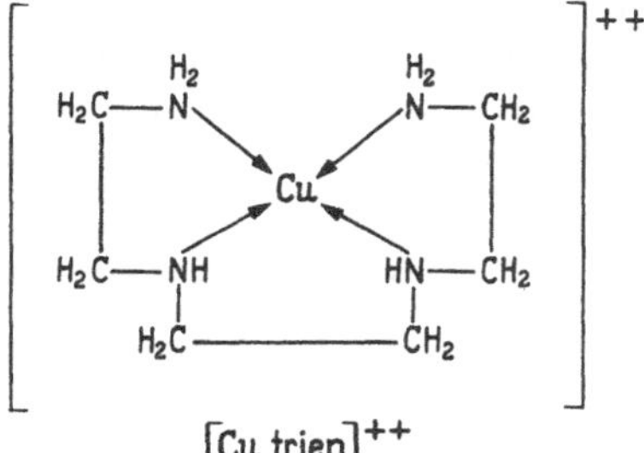

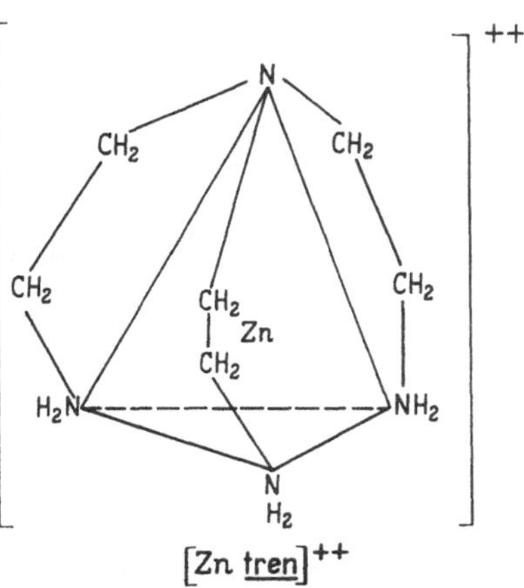

Fig. 1. Structure of some Typical Chelate Compounds

Table 3. *First Complex Formation Constants of Some Cations with Ammonia and Acetate*

Cation	$\log k_1$	
	NH_3	CH_3COO^-
Mg^{++}	—0.15[1]	0.5[2]
Ca^{++}	—0.2[3]	0.5[2]
Cd^{++}	2.54[1]	1.3[4]
Zn^{++}	2.37[3]	1.0[2]
Ni^{++}	2.77[5]	0.67[6]
Cu^{++}	4.14[5]	1.66[7]
Ag^{+}	3.20[3]	0.73[8]

c) Formation of Chelate Complexes

When a molecule carries several ligand groups which may coordinate simultaneously with a cation, the resulting complex is called a chelate. In MARTELL and CALVIN's

[1] DERR and VOSBURGH.
[2] CANNON and KIBRICK.
[3] BJERRUM (1941).
[4] LEDEN (1946).
[5] POULSEN and BJERRUM.
[6] FRONAEUS (1952).
[7] FRONAEUS (1951).
[8] MACDOUGALL and PETERSON.

nomenclature the chelate is said to be bidentate, tridentate, etc., depending on whether 2, 3 or more groups of any given complexing agent participate in chelate formation. Structural formulae of Cu^{++} chelates with two glycine molecules and with one molecule of triethylene tetramine *(trien)* and of Zn^{++} with triaminoethylamine *(tren)* are shown in Fig. 1.

It is generally found that whenever two ligand groups are carried by a molecule in such positions that complex ion formation results in 5 or 6 membered rings, combination with the bidentate reagent is favored over that with two monodentate complexing agents. The point may be illustrated on reactions of the type

$$M(NH_3)_2 + en \rightleftarrows M\,en + 2\,NH_3 \tag{8}$$

where *en* stands for the ethylene diamine. The thermodynamics of this reaction have been studied extensively by BJERRUM and NIELSEN; CALVIN and BAILAR; IRVING, WILLIAMS et al.; BASOLO and MURMANN (1952, 1954); SPIKE and PARRY; POULSEN and BJERRUM, and COTTON and HARRIS. It will be noted that the number of molecules increases as reaction (8) proceeds from left to right and an increase of entropy would, therefore, be expected to be the main driving force. The precise data obtained by SPIKE and PARRY and reproduced in Table 4 bear out this

Table 4. *Thermodynamic Quantities for the Displacement of Methylamine and Ammonia by Ethylenediamine*

Reaction	$\Delta F°$ kcal	$\Delta H°$ kcal	$\Delta S°$ e. u.
$[Cd(CH_3NH_2)_2]^{+2} + en$	—1.40	0	4.7
$[Cd(NH_3)_2]^{+2} + en$	—1.20	+0.1	3.7
$[Zn(NH_3)_2]^{+2} + en$	—1.55	+0.1	4.8
$[Cu(NH_3)_2]^{+2} + en$	—4.30	—2.6	5.7

expectation for complexes of Zn^{++} and Cd^{++} but in the case of Cu^{++} there appears to be also an important energetic preference for the *en* complex. Reaction (8) may be represented in two steps:

$$H_3NMNH_3 + NH_2CH_2CH_2NH_2 \rightleftarrows H_3NMNH_2CH_2CH_2NH_2 + NH_3 \tag{9}$$

$$H_3NMNH_2CH_2CH_2NH_2 \rightleftarrows M\begin{matrix} \diagup NH_2 \diagdown \\ \\ \diagdown NH_2 \diagup \end{matrix}(CH_2)_2 + NH_3 \tag{10}$$

and SCHWARZENBACH (1952) was the first to attempt an estimate of ΔS for (10) by assuming that the probability of ring closure is inversely proportional to the volume of a sphere whose radius is equal to the length of the linear metal-diamine complex. COTTON and HARRIS presented a more refined treatment of the entropy of ring closure and WESTHEIMER and INGRAHAM pointed out the analogy of (10) with the equilibrium

$$CH_3(CH_2)_n CH_3 \rightleftharpoons \begin{array}{l} CH_2 \\ | \quad \diagdown \\ | \quad \; (CH_2)_n + H_2 \\ | \quad \diagup \\ CH_2 \end{array} \tag{11}$$

In comparing (10b) and (11) it is necessary, however, to keep in mind that, in the case of the complex ion, changes in the orientation of solvent molecules and in the structure of the liquid will make additional contributions to ΔS. This point is emphasized by the observation that Cu^{++} forms a more stable complex with 2,2′-dimethyl-1,3-propanediamine than with 1,3-propanediamine (HARES, FERNELIUS and DOUGLAS) and by the smaller formation constant of copper divalinate as compared to the diglycinate (LI, WHITE and YOEST). In the first case the ligand is uncharged, while in the second case complex formation leads to charge neutralization. Both results are thus in agreement with the theory of FRANK and EVANS cited in the preceding section, according to which the "icebergs" formed in the vicinity of non-polar residues "melt" when constrained to the neighborhood of ionic charges.

The data collected in Table 5 illustrate the higher stability of 5-membered, as compared to 6-membered chelate rings. The outstanding exception is due to the requirement of a 180° angle between the

Table 5. *First Formation Constants of Chelates with 5- and 6-Membered Rings*

Cation	Chelating Agent	$\log K_{f_1}$	Reference
Cu^{++}	$H_2NCH_2CH_2NH_2$	10.89	POULSEN and BJERRUM
Cu^{++}	$H_2NCH_2CH_2CH_2NH_2$	10.07	POULSEN and BJERRUM
Cu^{++}	$H_2NCH_2COO^-$	8.62	MONK
Cu^{++}	$H_2NCH_2CH_2COO^-$	7.60	CURCHOD
Ni^{++}	$H_2NCH_2CH_2NH_2$	7.57	POULSEN and BJERRUM
Ni^{++}	$H_2NCH_2CH_2CH_2NH_2$	6.44	POULSEN and BJERRUM
Ni^{++}	$^-OOCCOO^-$	5.30	DENNEY and MONK
Ni^{++}	$^-OOCCH_2COO^-$	4.00	DENNEY and MONK
Ag^+	$H_2NCH_2CH_2NH_2$	4.70	SCHWARZENBACH, MEISSEN and ACKERMANN
Ag^+	$H_2NCH_2CH_2CH_2NH_2$	5.85	SCHWARZENBACH, MEISSEN and ACKERMANN

bonds coordinating donor atoms to Ag^+. COTTON and HARRIS showed that the normal preference for 5-membered rings is an entropy effect and that it is, in fact, slightly opposed by the lower potential energy of the 6-membered rings in which there is somewhat less repulsion between the hydrogen atoms. The entropy of ring closure may be made more positive by restraining the ligand groups of a chelating agent to a configuration favorable for the coordination of a cation. This principle was used by SCHWARZENBACH and ACKERMANN (1949) and by SCHWARZENBACH, GUT and ANDEREGG who compared *enta* with 1,2-diamino cyclohexane dicarboxylic acid

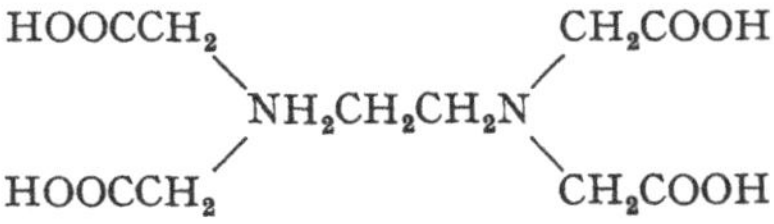

$$\begin{array}{ccc} HOOCCH_2 & & CH_2COOH \\ & \diagdown NH_2CH_2CH_2N \diagup & \\ & \diagup \qquad\qquad \diagdown & \\ HOOCCH_2 & & CH_2COOH \end{array}$$

enta

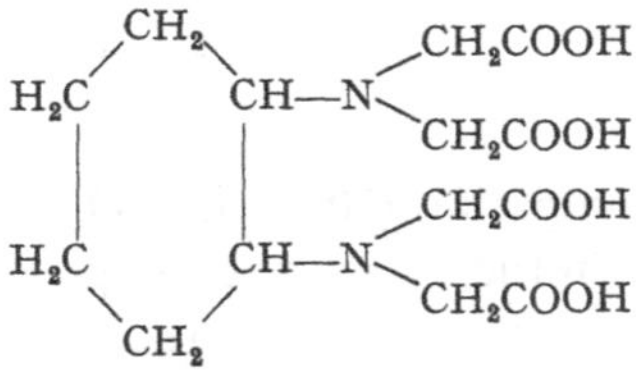

$$\begin{array}{ccccc} & CH_2 & & & CH_2COOH \\ H_2C & & CH & -N & \\ | & & | & & CH_2COOH \\ | & & | & & CH_2COOH \\ H_2C & & CH & -N & \\ & CH_2 & & & CH_2COOH \end{array}$$

1,2-diamino cyclohexane tetracetic acid

Complexes of the cyclohexane derivative were generally found to be more stable, as would be expected, but the magnitude of the effect showed a surprising dependence on the coordinated cation, with the ratio of the two formation constants varying from 600 for Cd^{++} to less than 2 for Ba^{++}. A similar demonstration of the effect of steric restraint on the stability and specificity of chelates was recently reported by GUTER and HAMMOND who found that dipivaloylmethane complexes specifically Li^{+} in the presence of other alkali ions.

The nature of the enthalpy change produced by the displacement of two complexed molecules by similar ligand groups of a bidentate chelating agent (IRVING, WILLIAMS et al.) are on the whole less clearly understood. WILLIAMS has pointed out that the equilibrium

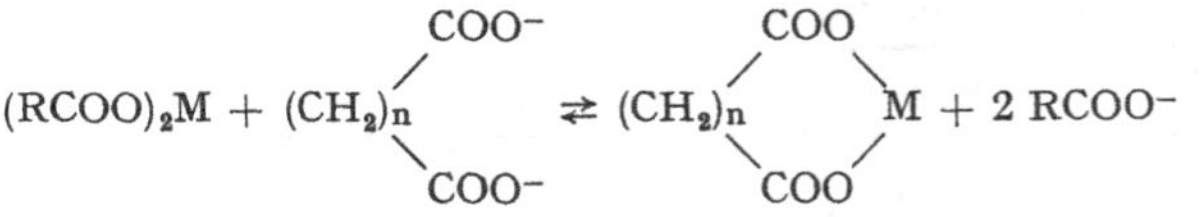

$$(RCOO)_2M + (CH_2)_n\begin{array}{l} \diagup COO^- \\ \diagdown COO^- \end{array} \rightleftarrows (CH_2)_n\begin{array}{l} \diagup COO \diagdown \\ \diagdown COO \diagup \end{array}M + 2\,RCOO^- \qquad (12)$$

should have a negative ΔH, since electrostatic potential energy is lost in the replacement of the doubly charged dicarboxylic acid anion by two monocarboxylic acid anions. Another important enthalpy effect is involved in the difficulty with which rings containing more than six members are generally formed. Little quantitative information is available on this point, but PEACOCK and JAMES report that the Cu^{++} malonate chelate has a formation constant 300 times as high as that of Cu^{++} succinate. This observation is quite in line with the dependence of ring closure probability on ring size, investigated for a number of chemical reactions (STOLL and ROUVÉ; BENNET; SPANAGEL and CAROTHERS). The rate of cyclization was found to drop sharply by a factor ranging from 5×10^4 to 4×10^{10} from a maximum for 5-membered rings to a minimum for rings containing 9—11 atoms. Larger rings were again formed more easily and STOLL and ROUVÉ found that the lactonization rate of α, ω, hydroxycarboxylic acids passed through a second maximum for 18-

membered rings, which formed 600 times more rapidly than 10-membered ones. STOLL and STOLL-COMTE interpreted this complicated behavior in terms of a steric interference of methylene hydrogens in rings of intermediate size and this will undoubtedly also govern the relative ease of ion chelate formation. It is hardly surprising that silver chelates with the 180° angle between the two coordinate bonds do not fit into the general pattern, so that the stability of cyclic complexes formed by Ag^+ with polymethylene diamines varies only very slowly with ring size (SCHWARZENBACH, MEISSEN and ACKERMANN).

If the ligand groups in a multidentate complexing agent are very widely spaced, the probability of ring closure becomes small and the ligands will be able to cooperate in complex formation only if the ΔH of the reaction is sufficiently negative. It is interesting to note in this context that LI, GAWRON and BASCUAS reported a stability constant of the Zn^{++} complex with oxidized glutathione which is explicable only by the participation of both terminal amine groups, implying the formation of an 18-membered ring.

A thorough analysis of systems containing Ag^+ and diamines has revealed the multiplicity of equilibria which have to be considered. For *en* ($pK_1 = 6.98$, $pK_2 = 9.98$), SCHWARZENBACH (1953) estimated the following equilibrium constants:

$$Ag^+ + H_2NCH_2CH_2NH_2 \rightleftarrows (AgNH_2CH_2CH_2NH_2)^+ \qquad \log K = 3.40 \qquad (13a)$$

$$^+AgNH_2CH_2CH_2NH_2 \rightleftarrows \left(\begin{array}{c} CH_2\text{——}CH_2 \\ | \qquad\quad | \\ H_2N\text{—}Ag\text{—}NH_2 \end{array}\right)^+ \qquad \log K = 1.3 \qquad (13b)$$

$$^+AgNH_2CH_2CH_2NH_2 + Ag^+ \rightleftarrows {}^+AgNH_2CH_2CH_2NH_2Ag^+ \qquad \log K = 2.55 \qquad (13c)$$

$$2\left[\begin{array}{c} CH_2\text{——}CH_2 \\ | \qquad\quad | \\ H_2N\text{—}Ag\text{—}NH_2 \end{array}\right]^+ \rightleftarrows \left[\begin{array}{ccc} & H_2 \qquad H_2 & \\ & N\text{—}Ag\text{—}N & \\ H_2C & & CH_2 \\ | & & | \\ H_2C & & CH_2 \\ & N\text{—}Ag\text{—}N & \\ & H_2 \qquad H_2 & \end{array}\right]^{++} \qquad \log K = 4.02 \qquad (13d)$$

$$Ag^+ + H_2NCH_2CH_2NH_3^+ \rightleftarrows (AgNH_2CH_2CH_2NH_3)^{++} \qquad \log K = 2.38 \qquad (13e)$$

As would be expected, increasing the distance between the ligand amine groups renders the dimer formation (13d) less probable, but favors the formation of the linear doubly charged species represented by (13c) and (13e). Similar complications would be expected in the complexation of the α, ω, dicarboxylic acid series which should be reinvestigated to provide a solid base for the interpretation of the chelation behavior of polymeric acids.

Many nucleophilic groups which by themselves would not show any detectable tendency towards complex ion formation will form very stable chelates in conjunction with a primary ligand group. Table 6 compares the equilibrium constants for the displacement of the phenolic proton by Fe^{+++} for phenol (MILBURN) and some of its o-substituted derivatives (AGREN) and shows that properly placed aldehyde, ketone, amide or ester groups may contribute to complex ion stability. SCHWARZENBACH and his collaborators (1955) have used in a similar manner the formation constants of N-substituted iminodiacetic acids, carrying various functional groups, to determine their relative coordination tendency. A particularly striking case was reported by SCHWARZENBACH, SENN and ANDEREGG (1957) who compared the Ca^{++} chelation with $(OOCCH_2)_2N(CH_2)_5N(CH_2CO\bar{O})_2$ and with $(^-OOCCH_2)_2N(CH_2)_2—O(CH_2)_2—N(CH_2COO^-)_2$. The substitution of an ether oxygen for a methylene group increased here the complex stability by a factor of 2.5×10^5.

Table 6. *Equilibrium Constants for* $Fe^{+++} + o\text{-}XC_6H_4OH \rightarrow o\text{-}XC_6H_4OFe^{++} + H^+$

o-Substituent	K
None	0.11
$—COOCH_3$	0.35
$—COCH_3$	0.5
—CHO	0.9
$—CONH_2$	13.5
$—COO^-$	500

ANDREWS has reviewed complex formation of aromatic hydrocarbons with cations, particularly Ag^+, and although these complexes have a rather low stability, a properly placed aromatic ring may be expected to increase appreciably the stability of complexes with compounds containing other ligand groups.

In some cases chelate formation stabilizes the enol form of a compound, as with diketones such as acetylacetone studied by VAN UITERT, FERNELIUS and DOUGLAS

$$CH_3\overset{O}{\overset{\|}{C}}—CH_2—\overset{O}{\overset{\|}{C}}—CH_3 + Cu^{++} \rightleftarrows (CH_3—\overset{O}{\overset{|}{C}}=CH—\overset{O}{\overset{\|}{C}}—CH_3)^+ + H^+ \quad \text{(O–Cu←O chelate ring)} \tag{14}$$

or amide derivatives of bases such as thiazole (v. HAHN, BÄUMLER, ROTH and ERLENMEYER)

$$\text{CONH}_2\text{-thiazole} + Cu^{++} \rightleftarrows \left[HN=C(—O—Cu \leftarrow N)\text{-thiazole} \right]^+ + H^+ \tag{15}$$

and it is particularly striking that a cation bound by a primary ligand group may displace protons of aliphatic hydroxyls to form a chelate, as in the Fe^{+++} complexes with citrate [WARNER and WEBER (1953a)] and gluconate (Pecsok and Sandera). On the other hand, in comparing the calcium complexation with citrate ($K_f = 1410$) and tricarballylate ($K_f = 66$) Schubert and Lindenbaum showed that coordination with the hydroxyl of the citrate stabilizes the complex, even though the hydroxylic proton is not being displaced in this case.

Mellor and Maley (1947, 1948) determined the order of the stabilities of bivalent cation complexes as

$$\mathrm{Pd} > \mathrm{Cu} > \mathrm{Ni} > \mathrm{Pb} > \mathrm{Co} > \mathrm{Zn} > \mathrm{Cd} > \mathrm{Fe} > \mathrm{Mn} > \mathrm{Mg}$$

and this order has been found to hold, in general, irrespective of the nature of the ligands. In aqueous solution only exceptionally strong chelating agents are able to complex alkali metal ions. Schwarzenbach, Kampitsch and Steiner (1946) found uramil diacetic acid to complex Li^+ ($K_f = 2.5 \times 10^5$) and Na^+ ($K_f = 2.1 \times 10^3$) and ethylenediamine tetracetic acid *(enta)* also binds these ions to a lesser extent [Schwarzenbach and Ackermann (1947)]. No K^+ binding was observed with these two complexing agents. Fernelius and van Uitert reported the same order for the alkali ion complexing by dibenzoylmethane in 75% dioxane. On the other hand, an ion exchanger has been described by Skogseid which is supposed to exhibit selective K^+ complexing characteristics (Craig).

Specific effects which may increase the affinity of a chelating agent for a given cation are of three types:

(1) Steric requirements of a chelate may favor some ion complexing agent combinations. Thus, data published by Prue and Schwarzenbach and by Schwarzenbach (1950) lead to an equilibrium constant of 10^4 for the chelate interchange reaction involving two of the complexes shown in Fig. 1:

$$\mathrm{Cu}\,\mathit{tren}^{++} + \mathrm{Zn}\,\mathit{trien}^{++} \rightleftarrows \mathrm{Zn}\,\mathit{tren}^{++} + \mathrm{Cu}\,\mathit{trien}^{++}\,.$$

This may easily be interpreted as due to the greater facility with which the linear *trien* may fit into the square planar Cu^{++} complex, while *tren* would be preferred in the tetrahedral Zn^{++} chelate.

Another example of the importance of the steric requirements for chelate formation is the high stability of dimethyl-glyoxime complexes interpreted by van Uitert and Fernelius as due to hydrogen-bond stabilization of the square planar structure, which may require a critical size of the central ion:

```
H3C—C———C—CH3
    ‖   ‖
    N   N
   ↙ \ / \
  O   \ /   O
  :   | /   H
  H   |/    :
  O    M    O
   \ ↗  \ ↗
    N    N
    ‖    ‖
H3C—C———C—CH3
```

(2) The inability of an ion to utilize all the ligand groups of a given chelating agent may reduce its relative affinity for that agent as compared to other ions. Thus, Cu^{++} binds the tetradentate chelating agent ethylenediaminediacetic acid 500 times more strongly than Ni^{++} while with ethylenediamine tetracetic acid *(enta)*, carrying six ligand groups which can all be utilized by Ni^{++}, but not by Cu^{++}, the binding of Ni^{++} appears, actually, to be slightly stronger (CHABEREK and MARTELL). In other cases, the size of an ion may determine the ease with which all the coordinate positions may be occupied by bulky ligand groups. The small Mg^{++}, which is bound eight times as strongly as Ca^{++} by the tridentate methyliminodiacetic acid [SCHWARZENBACH, KAMPITSCH and STEINER (1945a)] is held eighty times more weakly when the hexadentate *enta* is the chelating agent [SCHWARZENBACH and ACKERMANN (1947)].

(3) IRVING and WILLIAMS (1953) have pointed out that an electronic rearrangement may increase the stability of a chelate by a large factor. Outstanding cases of this effect are Fe^{++} complexes with three molecules of 2,2′-dipyridyl or o-phenanthroline *(phen)* in which the paramagnetism of Fe^{++} is lost. Thus, $\log K_1 K_2 K_3$ is 21.5 for $(Fe\ phen_3)^{++}$ as against 17.0 for $(Zn\ phen_3)^{++}$ and 15.2 for $(Cd\ phen_3)^{++}$, while for the paramagnetic $(Fe\ en_3)^{++}$ $\log K_1 K_2 K_3 = 9.53$, much smaller than the value of 12.09 for $(Zn\ en_3)^{++}$ and $(Cd\ en_3)^{++}$.

The formation of mixed chelates in which the cation forms a bridge between two different chelating agents has been observed in a number of cases [WATTERS and LOUGHRAN; DE WITT and WATTERS; JONASSEN, REEVES and SEGAL (1955a); SEGAL, JONASSEN and REEVES]. Complexes containing a chelating agent and hydroxyl coordinated simultaneously with the cation must often be taken into account. Treating chelates formally as acids which form hydroxo complexes by proton dissociation from a coordinated water molecule, SCHWARZENBACH and BIEDERMANN found pK values of 5 for $[Fe\ enta\ (OH)]^{--}$; 6.16 for $[Al\ enta\ (OH)]^{--}$; 7.5 for $[Cr\ enta\ (OH)]^{--}$; and JONASSEN, REEVES and SEGAL (1955b) obtained pK 13.3 for $[Cu\ en_2\ (OH)]^{+}$. Both the characteristic hydroxyl affinity of the cation and the nature of the chelating agent seem to play a part in these equilibria.

d) Ion Binding by Protein

Proteins participate in specific complex formation with both anions and cations and extensive reviews of this field have been presented by KLOTZ (1953), by SCATCHARD et al. (1954) and by GURD. With proteins as the complexing agents, some additional techniques may be employed with those used for the study of complexes of low molecular weight:

(1) The large size of the protein molecules makes it possible to use dialysis equilibrium as a convenient tool for determining the activities of diffusible species.

(2) As SØRENSEN, LINDERSTRØM-LANG and LUND have first pointed out, cations and anions are, in general, not bound in equivalent amounts, so that the net protein charge is changed and the isoelectric point (p_H of zero protein charge) no longer coincides with the isoionic point (p_H at which an equal number of cationic and anionic protein groups are ionized).

(3) Adsorbed cations increase, on electrostatic grounds, the tendency towards hydrogen ion dissociation and anion adsorption represses it; the addition of salts containing specifically bound ions to isoionic proteins will, therefore, produce a p_H shift from which the extent of ion binding may be calculated (SCATCHARD and BLACK).

(4) Changes in the catalytic activity of enzymes are particularly sensitive indications of the binding of activating or inhibiting ions to the site responsible for enzymatic activity.

The interpretation of experimental data is complicated by the necessity to take proper account of the effect of the protein charge on the association equilibria and by the multiplicity of groups which might be responsible for ion binding. The problem is analogous to the titration of a weak polyelectrolyte as discussed by KATCHALSKY and GILLIS and with N association sites of a single type characterized by an intrinsic association constant k^0, the average $\bar{n}$ of ions bound to a protein molecule is

$$\bar{n} = \frac{N(A)\, k^0 \exp(-\Delta F_{el}/RT)}{1 + (A)\, k^0 \exp(-\Delta F_{el}/RT)} \tag{16}$$

where (A) is the activity of the free ions, ΔF_{el} being the electrostatic free energy of ion binding. For spherical isoionic protein molecules of radius b, SCATCHARD (1949) gives

$$\Delta F_{el} = \frac{(2\,i-1)}{2}\,\frac{\bar{N} Z^2 e^2}{D}\left[\frac{1}{b} - \frac{\varkappa}{1+\varkappa a}\right] \tag{17}$$

where i is the number of small ions of charge Z bound to the protein, $\bar{N}$ is AVOGADRO's number, e the electronic charge, D the dielectric constant, a the closest distance of ion-protein approach and $\varkappa$ the Debye-Hückel parameter. Equation (16) may be arranged to

$$\frac{\bar{n}}{(A)} \exp(\Delta F_{el}/RT) = k^0 N - k^0 \bar{n} \tag{18}$$

a form in which it is used conveniently for estimating k^0 and N from a plot of experimental data.

STEINHARDT (1941, 1942) and STEINHARDT, FUGITT and HARRIS were the first to note the differences in the titration curves of wool or egg albumin with various strong acids and they interpreted their data as evidence of anion binding to the proteins. For wool, the p_H at half neutralization was 2.33 with HCl, 2.58 with HNO_3, 3.08 with H_2SO_4 and

3.86 with picric acid. These data, as well as a study by SCATCHARD and BLACK of the p_H shift of isoionic serum albumin when various salts were added, indicated that the more polarizable anions, or those carrying bulky hydrocarbon residues have the highest affinities for proteins. The binding of such ions as dodecyl sulphate or methyl orange may involve, as a contributing factor, a similar interaction of hydrophobic groups as produces soap micelle formation. However, the high affinities reported by SCATCHARD, SCHEINBERG and ARMSTRONG for serum albumin binding of ions such as chloride ($k^0 = 44$) and thiocyanate ($k^0 = 1000$) is difficult to understand. KLOTZ and URQUHART have pointed out that the ability of serum albumin to bind anions is much stronger than that of other proteins and found that it decreases with acetylation but remains unchanged on guanidation of the ε-amino groups. They ruled out the possibility that the cooperation of several groups may account for anion binding and interpreted their data in terms of anion association with such cationic sites of the protein as are not inactivated by ion pair formation with carboxylate groups. While no alternative suggestion for an interpretation of anion binding appears to have been offered, the interpretation of KLOTZ and URQUHART would imply anion association with simple aliphatic amines which has not been observed.

The activation of enzymes by anions is usually due to the removal of an inhibitory cation by complex formation, but the activation of amylase by chloride ion is undoubtedly due to chloride association with the enzyme molecule. The association constant must be quite high since MYRBÄCK (1926) found that 5×10^4 N chloride produces half of the full activation. However, MICHAELIS and PECHSTEIN could interpret their data only by postulating that the affinity of a given anion for the enzyme and the catalytic activity of the resulting complex are two unrelated parameters, with nitrate forming the most stable complexes and chloride the most active ones.

While a firm basis is still lacking for the understanding of anion binding, the formation of cation complexes with proteins is understandable, at least in principle, in terms of cation interactions with small molecules containing similar functional groups. Ferritin presents a special case; it contains variable amounts of iron — up to 23% by weight — apparently in the form of a basic Fe^{+++} phosphate micelle imbedded into the globular protein molecule (GRANICK). The heme proteins (WYMAN) contain Fe^{++} linked to the four nitrogen atoms of a protoporphyrin ring, and the two remaining coordinate positions of the cation provide, presumably, points of attachment to ligand groups of the protein. In the case of haemoglobin the imidazole groups of histidine residues have been postulated as the combining sites, and the recent observation by CORWIN and REYES that the imidazole complex of ferroprotoporphyrin can form reversible complexes with molecular oxygen is strong support for this theory. On the other hand, hemocyanin, the respiratory protein found in many crustaceans and mollusks, contains no prosthetic group and has copper linked directly to functional groups of the protein moiety (HAUROWITZ and HARDIN; DAWSON and MALLETTE). It is the only copper compound known to combine reversibly with oxygen and the clarification of the type of complex which will endow Cu^{++} with this property remains a challenge to the investigator. A similar mystery shrouds the manner of vanadium binding in hemovanadin [BIELIG and BAYER (1953, 1954)], found in the strongly acidic blood of ascidians. The vanadium concentration in the blood corpuscles of these organisms is 3×10^6 times as high as in the surrounding sea water, a truly remarkable example of the efficiency and selectivity of complex formation attainable in biological systems.

In general, the nature of the groups involved in the binding of cations to proteins is inferred from the following type of evidence:

(1) A comparison of the intrinsic association constant k^0 of a cation with a protein and a suitably chosen low molecular weight analog.

(2) The correspondence of the apparent number of sites available for the binding of a given cation with a similar number of a given type of groups known to be present in the protein.

(3) Changes of the complexation pattern after chemical modification of the protein.

(4) Spectroscopic evidence.

Strong ligand groups are provided by terminal carboxyl and amine, the β and γ carboxyls of aspartic and glutamic acid residues, the ε-amino groups of lysine, the thiol of cysteine, the imidazole of histidine and phenolic hydroxyls of tyrosine. The hydroxyl of serine and the disulfide bonds of cystine may be involved in chelate formation. Finally, DOBBIE, KERMACK and LEES; DATTA and RABIN (1956a, b), and MANYCK, MURPHY and MARTELL have recently shown that the peptide group, whose contribution to cation binding has been ignored by most workers in this field, has a profound effect on the stability of Cu^{++} complexes with peptides. With glycylglycine the stabilization of the primary complex by chelation with the ionized peptide group has been represented by

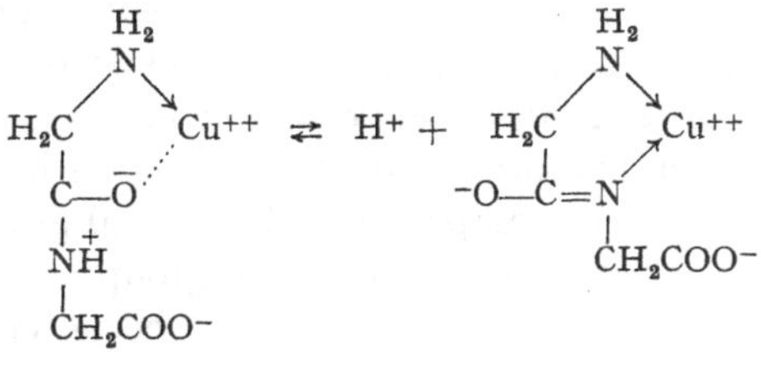

$$\begin{array}{ccc} & H_2 & \\ & N & \\ H_2C & & Cu^{++} \\ | & & \\ C{-}\bar{O} & & \\ |\,+ & & \\ NH & & \\ | & & \\ CH_2COO^- & & \end{array} \rightleftarrows H^+ + \begin{array}{ccc} & H_2 & \\ & N & \\ H_2C & & Cu^{++} \\ | & & \\ {}^-O{-}C{=}N & & \\ & | & \\ & CH_2COO^- & \end{array} \qquad (19)$$

with a pK of 4.25 and complex formation with diglycylglycine was shown to involve the ionization of both peptide bonds with pK values of 5.2 and 7.0. In the "biuret complex" of Cu^{++} with proteins in strongly basic solution, both the composition of the complex determined by MEHL et al. and spectroscopic comparison with a series of glycine polypeptide complexes (PLEKHAN) points to Cu^{++} coordination with four ionized peptide groups.

Cation binding to serum albumin, a globular protein in which the folding of the polypeptide chain should lead to a well defined steric relationship of the various ligand groups, has been the object of extensive investigations. KLOTZ and CURME studied the binding of Cu^{++} to serum albumin and interpreted equilibrium and spectral data as pointing to 16 binding sites consisting of carboxyl in suitable juxtaposition with other ligand groups. TANFORD (1952) studied complex formation of Cu^{++}, Zn^{++}, Cd^{++} and Pb^{++} by polarography and concluded from the p_H range over which the complex dissociated that the cations formed complexes with imidazole residues. Spectral evidence was also cited in favor of this conclusion, although the spectrum of the Cu^{++} complex with a single imidazole group has not been determined and the absorption maximum of $[Cu(NH_3)]^{++}$ lies at 745 mμ (BJERRUM, BALLHAUSEN and JØRGENSEN) high above the peak at 670 mμ reported by KLOTZ, FULLER and URQUHART for Cu^{++} in serum albumin solution at p_H 7. GURD and GOODMAN studied the binding of Zn^{++} by serum albumin and interpreted their data also in terms of cation binding to imidazole groups of histidine residues. The calculated association constant was found, in fact, to lie close to that determined by EDSALL et al. for Zn^{++} association with imidazole. The shape of the association isotherms given by GURD and GOODMAN does not appear, however, to be consistent with a single type of binding site. HUGHES and KLOTZ have also found that the capacity of acetylated serum albumin for Zn^{++} rises continuously with p_H well beyond the range of the ionization of imidazole groups and attains values in excess of the histidine content of the protein. These discrepancies may be due at least in part to the role of peptide bonds in stabilizing cation complexes with carboxylate and other groups (RABIN).

The cooperation of different protein groupings has been postulated in several cases to account for cation binding. WARNER and WEBER (1953b, 1954) studied the Fe^{+++} complex of conalbumin, which absorbs strongly at 440 mμ, suggesting by analogy with known Fe^{+++} complexes the involvement of a phenoxide group. The disappearance of the absorption band attending protein denaturation was interpreted as due to the participation, in the complex, of other ligand groups, which have the necessary steric relation to the phenolic group when the protein is in its native state, but are too far apart in the polypeptide chain to cooperate in chelate formation when the protein is denatured. A similar cooperative effect was postulated by KLOTZ et al. (1955) who cite spectroscopic evidence for the existence of two different types of binding sites when bovine serum albumin associates with Cu^{++} ions. One of these sites is characterized by a strong absorption band at 375 mμ and the effect of various reagents on light absorption at this wavelength led to the conclusion that this band is due to a Cu^{++}-thiol complex stabilized by coordination of a properly spaced disulfide group. KLOTZ and his collaborators (1957) have also presented convincing evidence to clarify the binding of iron to hemerythrin, the respiratory protein of marine worms. Spectroscopic data point to the attachment of Fe^{++} to thiol groups and the fact that Hg^{++} has twice the inactivating effect of Ag^{+} was interpreted to signify that the Fe^{++} involved in oxygen transport forms a bridge between two closely spaced cysteine residues.

The extremely high value of the association constant of Hg^{++} with mercaptans may lead to the dimerization of proteins containing a single reactive thiol group when one mol of mercuric salt is added to a solution of two mols of protein. This striking phenomenon was first discovered by HUGHES (1947, 1950) on a fraction of serum albumin (mercaptalbumin) and the thermodynamics and kinetics of the process were subjects of an intensive study (EDELHOCH et al.). More recently, KIMMEL and SMITH and SMITH, KIMMEL and BROWN demonstrated a similar behavior on papain.

LEHNINGER has stressed in his review the high specificity of the activator requirement in cation activated enzymes. In some cases the specificity is absolute, as with carbonic anhydrase (VALLEE) requiring Zn^{++} or prolidase [SMITH and BERGMANN, SMITH et al. (1954b)] activated only by Mn^{++}. It is noteworthy that cations known to form very stable complexes, such as Cu^{++}, function as activators only in a few enzyme systems (SINGER and KEARNEY), while the very weakly complexing Mg^{++} and Mn^{++} are most often found to be specifically required. It is unexpected to find that enzymes catalyzing the same reaction but derived from different biological sources may be activated by different cations [SMITH (1948)]. The ion required for activation has been used frequently to differentiate between enzymes, but a recent report by VESCIA, which should await confirmation, claims that the substrate specificity of a single enzyme may depend to some extent on the activating ion. MALMSTRÖM (1955) has stressed the fact that enolase is activated both by Zn^{++} which is known to form stable complexes and by Mg^{++} which is a very weak complex former, while cations intermediate in their complexation behavior are inhibitory. The size of the ion bound to the enzymatically active site was assumed to be here a critical factor. KACHMAR and BOYER used a similar argument to account for the fact that pyruvic phosphoferase required in addition to Mg^{++} the presence of K^+, Rb^+ or NH_4^+, but was inactive if the only alkali ion present was Na^+ or Li^+. WEBSTER and VARNER reported studies on a glutamyl cysteine synthetizing enzyme which required Mg^{++} and K^+ with no other alkali ion being able to substitute for potassium. These findings are particularly striking, in view of the fact that Na^+ generally exceeds K^+ in its tendency towards complex formation.

With some enzymes, the removal of the activating ion from the enzyme produces irreversible enzyme inactivation (VALLEE) while in a number of other cases the formation of the active complex is reversible. MALMSTRÖM has stressed the fact the apparent enzyme-activator association constant will depend on the complex forming properties of the buffer used in the enzyme assay. He reports an association constant of 3300 for Mg^{++} with enolase in tris buffer at p_H 7.2 and a comparison of this value with the low Mg^{++} association constants with simple ligand groups (Table 3) is proof that several groups must cooperate in the enzymatically active site to result in such tight binding. It is instructive to point out that on the basis of data by SCHWARZENBACH, KAMPITSCH and STEINER (1945a) the binding of Mg^{++} to the catalytically active site of enolase appears to be as strong as that to the tetradentate chelating agent ammonia triacetic acid at the same p_H. The activating ion may also stabilize the enzyme against denaturation [SMITH et al. (1954b)], a clear indication that the groups engaged in chelate formation are close to each other only when the polypeptide chain of the enzyme is folded in the highly specific manner characterizing the native state. COHN, and COHN and TOWNSEND have recently pointed out that paramagnetic resonance absorption may be used to determine the extent of complex formation of paramagnetic ions (such as the important enzyme activator Mn^{++}) with solution samples of a fraction of 1 ml. This technique may allow comparisons of the effect of added ions on catalytic activity with the enzyme-activator association equilibrium, which could be determined previously only in rare cases [MALMSTRÖM (1953)] because of the small size of available enzyme preparations.

An ion, which in low concentration activates an enzyme, sometimes becomes an inhibitor at higher concentrations [MALMSTRÖM (1953), RABIN and CROOK]. This behavior may be interpreted in terms of two cation binding sites, one of which must be complexed and the other free for the enzyme to be catalytically active. The phenomenon is strictly analogous to the typical dependence of enzyme activity on hydrogen ion activity as analyzed in detail by WALEY, by DIXON and by ALBERTY.

The activating ion seems to fulfill different roles in different enzymatic reactions [KLOTZ (1954)]. STEINBERGER and WESTHEIMER have shown that in some cases the activating ion has, by itself, some catalytic action and complex formation with ligand groups of the protein may then only serve to enhance the catalytic effect. KLOTZ (1954) has pointed out that complexation with the apoenzyme will keep in solution catalytically active ions which would tend to be hydrolyzed at physiological p_H. A particularly striking demonstration of the manner in which the catalytic property of an ion may depend on a highly specific chelate form is due to WANG. He showed that the catalysis of hydrogen peroxide decomposition by Fe^{++} is enormously increased when the ion is bound to four nitrogens of triethylenetetramine, while diethylenetriamine or tetraethylene pentamine complexes have very slight activity. In the case of some cation activated peptidases, SMITH (1941) has proposed that the activating ion may be chelated to both enzyme and substrate, acting as a bridge in the enzyme-substrate association complex. KLOTZ and LOH MING have modified this picture to account for the frequency with which weakly complexing cations are specifically required activators. They stress the fact that the cation must serve to stabilize the activated complex in the substrate decomposition rather than the ground state of the enzyme-substrate association complex. Studies of model systems in which complex forming cations formed mixed complexes with a protein and a small dye molecule showed that the effectiveness in this "mediating" action of an ion does not necessarily parallel its general affinity for ligands. This was ascribed to the probability that weaker ligand groups of the protein would saturate strongly complexing cations while less reactive cations, bound only to one or two strong ligand groups of the protein, are left free to coordinate also with the dye. The recent observation by DATTA and RABIN that the peptide hydrogen of diglycine is displaced by Cu^{++} but not by Co^{++} or Mn^{++} appears to be strong supporting evidence for this interpretation.

The heavy metal inhibition of enzymatic activity shows usually little specificity but follows the normal sequence of the affinity of the cations for ligand groups. Enzymes whose activity depends on a free thiol group (BARRON) are naturally most sensitive to such inhibition. AMBROSE, KISTIAKOWSKI and KRIDL followed the dependence of urease inhibition on the concentration of free silver ion and found that as little as 10^{-10} M Ag^+ sufficed to reduce the enzyme activity by 50%. The relative effectiveness of inhibiting cations may be assessed from results obtained by MYRBÄCK (1955) with saccharase which formed complexes with Ag^+, Cu^{++}, Cd^{++} and Zn^{++} characterized by dissociation constants of 9×10^{-8}, 10^{-6}, 3×10^{-6} and 3×10^{-5}, respectively.

e) Specific Ion Binding by Extended Chain Molecules

The difference of the problems encountered in studying ion binding to proteins and linear chain polyelectrolytes may be illustrated by a comparison of serum albumin and polymethacrylic acid. Human serum albumin is a globular protein whose molecular shape changes little with the ionic charge. Since it is amphoteric, ion binding may be studied at the isoelectric point, where all complications due to electrical charge effects disappear. But even in moving from the isoelectric p_H of 5.3 to $p_H = 7$ the protein charge is only -13 or one negative charge per 5000 molecular weight units [TANFORD (1950)]. Polymethacrylic acid, on the other hand, has flexible chain molecules which tend to extend greatly with increasing degree of ionization. The charge density per molecular weight unit may attain very large values; in salt solution at p_H 7 it is about 60 times as high as that of the serum albumin.

Under these conditions $\varepsilon\,\psi$, the product of the electronic charge and the electrostatic potential, is not small compared to kT as required by the electrolyte theory of DEBYE and HÜCKEL and the concept of "ionic strength" becomes useless. At a given concentration, the mean activity coefficient of a salt with divalent cations and monovalent anions will no longer be expected to have the same mean activity coefficient as one with monovalent cations and divalent anions, since the valency of the counter-ions of the polyelectrolyte will affect the electrostatic forces in the solution much more than the valency of the by-ions. In addition to long range electrostatic interaction, OSAWA et al., WALL et al. (1956) and RICE have postulated the existence of ion-pairs within the highly charged polymer coil. These theories are supported by the data of FUOSS and CHU whose conductance measurements indicate that bis-quaternary ammonium salts form ion pairs stable even in aqueous solution. On the other hand, MEEKS and MARCUS, who investigated the dependence of the solubility of silver succinate and sebacate on the concentration of added sodium nitrate, could account for their data by a modification of the DEBYE-HÜCKEL treatment without assuming ion pair formation.

True complex formation differs from these non-specific interactions in the conditions imposed by the coordination number and the steric requirements of the complexing cation, the characteristic relatively large difference in the behavior of different ions and the important role in complex stabilization frequently played by uncharged ligand groups.

The multiplicity of possible interactions of polyelectrolytes with small ions would lead one to expect that results in the determination of "ion binding" would depend on the method chosen for its determination. The elegant tracer technique used by HUIZENGA et al. (1950a, b) and by WALL et al. (1952) to determine the association of counter ions with the macro-ion in electrophoresis and diffusion should demonstrate a composite effect due to all the causes discussed above. The unusually rapid increase in the conductivity of polyelectrolyte solutions with increasing electrical field strength (Wien effect) reported by BAILEY,

Patterson and Fuoss and the dependence of the transference number on field strength [Wall et al. (1956)] indicate the range of degrees of counter-ion binding. A useful limiting case is that of polyelectrolyte solutions containing small concentrations of complex forming ions in the presence of a large excess of salts composed of ions which are known not to participate in complex formation (e. g. K^+, NO_3^-, ClO_4^-). In such a system electrostatic interactions with the polymer or ion-pair formation with its ionized groups will have a negligible effect on the activity of the simple electrolytes and the activfty of the complex forming species may be interpreted in terms of complexation with the polyelectrolyte. The tracer technique of Huizenga et al. and of Wall et al. could presumably also be used to prove complex ion formation, since only the transport of a complexed ion with the polymer will persist in the presence of large excess of non-complexing electrolytes.

The addition of neutral salts to polyelectrolyte solutions containing weakly ionizing groups will generally lead to changes of p_H due to changes in the electrical free energy of ionization. However, once a sufficient concentration of neutral salt is present, this effect becomes negligible for further small salt additions and any observed p_H shift may be interpreted as due to competition between the added cation and hydrogen ion for ligand groups on the polyion. The salt used to swamp electrostatic field effects of the polyelectrolyte should, of course, contain no complex forming ions.

In applying the Bjerrum (1941) procedure to the interpretation of the p_H shift obtained when polymeric acids carrying a single type of ionizing group participate in complex formation, account must be taken of the fact that even in strong salt solution the apparent ionization constant K_a of a polyion varies with the degree of ionization. The complexation of counter-ions will also affect K_a but the extent of this change cannot be determined unambiguously. Neglecting specific effects of counter-ion chelation on the configuration of the polyion and possible neighboring group effects, Morawetz, Kotliar and Mark assumed that K_a depends only on the net charge density Z along the polymeric chain. Equations (3) must then be replaced, for the binding of bivalent cations, by the relations

$$\begin{aligned}
(\mathrm{A}) &= C_A\,\alpha + (\mathrm{H}) - (\mathrm{OH}) - (X)\\
C_A &= (\mathrm{HA}) + (\mathrm{A}) + (X)\\
\frac{(\mathrm{H})\,(\mathrm{A})}{(\mathrm{HA})} &= K_a(Z)\\
Z &= (\mathrm{A}) + \frac{\bar{n}-2}{\bar{n}}\,\frac{(X)}{C_A}
\end{aligned} \qquad (20)$$

which may be combined to

$$\frac{2\,K_a(Z)}{n\,Z\,C_A + (2 - \bar{n})\,[\alpha\,C_A + (H) - (\mathrm{OH})]} = \frac{(\mathrm{H})}{C_A(1-\alpha) - (\mathrm{H}) + (\mathrm{OH})}\,. \tag{21}$$

The function $K_a(Z)$ is obtained from the titration of the polymeric acid in the absence of complexing ions and equation (21) may be evaluated provided $\bar{n}$ is known. In the case of the binding of alkaline earth cations by hydrolyzed maleic anhydride copolymers, studied by MORAWETZ, KOTLIAR and MARK, it could be assumed that a chelate involving two neighboring carboxyl groups is the only type of complex formed and the titration shift could be interpreted in terms of the extent of counter-ion binding. These systems represent a particularly simple case of chains carrying well defined chelate forming groups of ligands, and the results may be compared with the chelate formation constants of a low molecular analog (i. e. succinate) with superimposed effects due to the electrostatic field of the polyion or changes in the adjoining medium of the chelates as the polyion stretches out with increasing net charge.

The binding of Cu^{++} to polyacrylic and polymethacrylic acids, which have been investigated by WALL and GILL, by GREGOR, LUTTINGER and LOEBL and by KOTLIAR and MORAWETZ, presents much more difficult problems. The high affinity of the polyanions for Cu^{++} leaves no doubt that chelate formation must be involved. The cooperation of two neighboring carboxylates would, however, result in eight-membered rings which are known to form with great difficulty (STOLL and ROUVÉ, SPANAGEL and CAROTHERS) and larger rings involving carboxylates at longer distances along the polymeric chain, must account for most of the cation binding. FRONAEUS' (1951) data on the binding of acetate by Cu^{++} also make it clear that the possibility of association with more than two carboxylate groups cannot be disregarded when a Cu^{++} ion finds itself surrounded by the high local concentration of carboxylate in the region occupied by the polyanion.

In analyzing the factors which determine the equilibrium involving a polyanion and a complex forming cation, it is convenient to make a fundamental distinction between the first step, in which the free cation combines with a single ligand of the polymer, and subsequent steps in which the bound cation forms chelates by coordination of further ligand groups. The equilibrium of the first step will depend on the stoichiometric concentration of ligand groups in solution and will also be affected by the electrostatic field of the polyion,

$$\frac{(\mathrm{CuA})^+}{(\mathrm{Cu}^{++})\,(\mathrm{A})} = K_1^0 \exp(-\Delta F'_{el}/R\,T) \tag{22}$$

where $\Delta F'_{el}$ is the electrostatic free energy corresponding to the first association step and K_1^0 is the first complex formation constant of a suitable monodentate analog. Assuming that electrostatic factors account for all of the variation of the apparent ionization constant of polymeric acids with the degree of ionization, $\Delta F'_{el}$ is related to the electrostatic free energy of ionization by

$$\Delta F'_{el} = -2\,\Delta F_{el} \tag{23}$$

and we obtain

$$\frac{(\mathrm{CuA})^+}{(\mathrm{Cu}^{++})\,(\mathrm{A})} = K_1^0\left(\frac{K_a^0}{K_a}\right)^2 \tag{24}$$

where K_a and K_a^0 are the apparent and the intrinsic ionization constants, respectively.

In dilute solutions of polymeric acids and at a high excess of carboxyl over Cu^{++}, the probability of chelation with two carboxylate groups of the same chain will be heavily favored over the association with groups belonging to different polymers. Such chelation equilibria depend, therefore, only on the state of the isolated macromolecular coils, being independent of their number as long as interpenetration of the polymers can be neglected. We have then equilibria of the form

$$\frac{(\mathrm{CuA_2})}{(\mathrm{CuA})^+} = K_2'\,; \qquad \frac{(\mathrm{CuA_3})^-}{(\mathrm{CuA_2})} = K_3'\,; \qquad \ldots \tag{25}$$

where K_2', K_3' etc. depend only on the degree of ionization of the polyion, provided the number of ligand groups is very large compared to the number of chelated cations so that interference between different parts of the polymer engaged in chelate formation may be disregarded. At relatively higher "loading" of the polyion with bound cations, such interference will result in a severe restraint on possible chain configurations and the magnitude of K_2', K_3', etc., would be expected to decline.

In view of the factors described above, the calculation of "chelate formation constants", in the conventional sense, has no meaning in the chelation of ions by polyelectrolytes. The best approach to a study of such systems requires the use of at least two experimental methods, one to determine the concentration of the free ions and the other to provide information about the type of complex formed. Using for the concentration of all species of bound copper

$$\mathrm{Cu_b} = \Sigma_i(\mathrm{CuA_i}) \tag{26}$$

equation (24) and (25) may be combined to

$$\begin{aligned} \frac{\mathrm{Cu_b}}{(\mathrm{Cu})^{++}} &= \overline{K}(\mathrm{A}) \\ \overline{K} &= K_1^0\left(\frac{K_a^0}{K_a}\right)^2 [1 + K_2' + K_2'K_3' + \cdots]\,. \end{aligned} \tag{27}$$

The ratio of $Cu_b/(Cu^{++})$ was determined polarographically by Wall and Gill and by dialysis equilibrium by Kotliar and Morawetz. With polyacrylic acid, both methods gave $Cu_b/(Cu^{++})$ proportional to the first power of (A), proving that any but the first cation association step is unimolecular. With polymethacrylic acid, dialysis equilibria indicated a decrease in the affinity of the polyion for Cu^{++} with increasing polyelectrolyte concentration, probably due to molecular association of the polyanion.

Spectroscopic studies both in the visible (KOTLIAR and MORAWETZ) and the ultraviolet (WALL and GILL) led to the surprising conclusion that the degree of ionization of the polymeric acid affected only the extent of copper binding but not the nature of the chelate formed. The very high intensity of the ultraviolet absorption band ($\varepsilon \sim 4500$ at 252 mμ and 257 mμ for Cu^{++} in half-neutralized polymethacrylic and polyacrylic acid, respectively) would recommend further studies of ultraviolet spectra of polyelectrolyte solutions containing complex forming ions. A comparison of Cu^{++} spectra in solutions of polymeric acids, acetate and malonate [MORAWETZ (1955), ROSENHECK] seems to indicate that four carboxylate groups of the polymeric acids participate in Cu^{++} chelation.

The principles discussed above for cation complex formation with polymeric acids apply also for corresponding complexes with polymeric bases or polyampholytes. GREGOR has reported cation binding studies with polyethylene imine, polyvinyl imidazole and poly-N-ethyleneglycine and MORAWETZ and SAMMAK have studied Cu^{++} binding by poly(ε-methacrylyl lysine). In all these cases the high local concentration of ligand groups in the swollen polymer coil favors the formation of the higher chelates. For instance, spectroscopic data obtained by MORAWETZ and SAMMAK show that in a solution of poly(ε-methacrylyl-l-lysine) containing 0.024 moles of monomer units per liter, the ratio of copper chelates with two and with one aminoacid group, respectively, is as high as in a 0.8 M solution of simple aminoacid. CORNAZ and DEUEL have described a polyhydroxamic acid which formed very stable complexes with Fe^{+++} and although their material was cross-linked, there is no doubt that a linear polymer with hydroxamic acid groups will exhibit the same specificity.

The precipitation of polymeric acids on addition of divalent cations [DEUEL and SOLMS (1951), WALL and DRENAN] may, but need not, be an indication of counter-ion binding. Both polyacrylic acid and polymethacrylic acid are precipitated by calcium chloride in neutral solution, but KOTLIAR found that only in the case of polyacrylic acid is the titration curve at high ionic strength shifted by Ca^{++}, indicating complex formation. On the other hand, hydrolyzed maleic anhydride-vinyl alkyl ether copolymers are not precipitated by such ions as Cu^{++} (SEYMOUR, HARRIS and BRANUM) although according to KOTLIAR they form very stable chelates. A particularly interesting case is that of carrageenin, a polysaccharide carrying sulphate monoester groups, which is extracted from a red alga. Carrageenin contains a fraction which is precipitated or gelled, depending on conditions, by K^+, Rb^+, Cs^+ and NH_4^+ but not by Na^+ and Li^+ (SMITH and COOK). O'NEILL and SMITH, O'NEILL and PERLIN found that the potassium sensitive fraction ($\varkappa$-carrageenin) contains 3,6-anhydro-d-galactose residues, which are absent in the fraction which is soluble in the presence of potassium salts. BAYLEY postulated that complex formation depending on the smaller radius of hydrated K^+, Rb^+ etc., as compared to Na^+ and Li^+ accounts for the phenomenon, but SCHACHAT and MORAWETZ found no evidence of selective K^+ binding in dialysis equilibrium studies. Preferential K^+ complexing, as pointed out above, would run counter to the relative stability of alkali ion complexes with well defined chelating agents but would resemble the behavior of K^+ activated enzymes. Precipitation of a colloid by electrolytes may be due to a "salting out" process, which has nothing in common with ion binding by the polymer. When complex formation does occur it may, or may not, lead to precipitation, depending on the general solvation of the polymers[1]. The same points may be made with respect to reversible gel formation of such substances as pectin and alginic acid in the presence of multivalent cations. The cross-linkages of the gel structure seem to be due to hydrogen bonding between the hydroxyl groups of the sugar residues and the ability of the polymer to gel is thus highly sensitive to the steric distribution of these groups (DEUEL, SOLMS and ALTERMATT). It is also significant that DEUEL and SOLMS (1954) found the gelation of pectin with Ca^{++} to be prevented much more rapidly by acetylation of hydroxyls than by methylation of the carboxyl groups.

A very important observation was made by STRAUSS, GERSHFELD and SPIERA, who found that quaternized poly (vinyl pyridine) may acquire a negative charge in solutions of high bromide concentration. This proves that the cationic groups of the polymer can bind more than an equivalent amount of counter-ions and that other than long range electrostatic forces must be involved. This must also be the case in the pronounced anion binding to the unionized polymer poly(vinyl pyrrolidone) described by FRANK, BARKIN and EIRICH. These phenomena have an obvious bearing on the nature of anion binding to proteins and it is to be hoped that future studies with model systems containing groups characteristic of proteins will lead to a further clarification of this important problem.

---

[1] A similar observation has recently been made in connection with a study of counter-ion binding to polyphosphate [U. P. STRAUSS, D. WOODSIDE and P. WINEMAN, J. Phys. Chem., **61**, 1353 (1957)]. Although electrophoretic mobility data show that Na^+ and Li^+ are equally bound, the polymer is precipitated much more rapidly by Na^+ salts. This discrepancy was interpreted as signifying that bound Li^+ is more highly solvated than bound Na^+.

Bibliography

AGREN, A. (1954): Acta chem. Scand. **8**, 266, 1059.

— (1955): Acta chem. Scand. **9**, 39, 49.

ALBERTY, R. A. (1954): J. Amer. chem. Soc. **76**, 2494.

AMBROSE, J. F., G. B. KISTIAKOWSKI and A. G. KRIDL (1951): J. Amer. chem. Soc. **73**, 1232.

ANDREWS, L. J. (1954): Chem. Rev. **54**, 713.

BAILEY, F. H., A. PATTERSON JR. and R. M. FUOSS (1952): J. Amer. chem. Soc. **74**, 1845.

BALLHAUSEN, C. J., and C. K. JØRGENSEN (1955): Acta chem. scand. **9**, 397.

BARRON, E. S. G. (1951): Advanc. Enzymol. **11**, 232.

BASOLO, F., and R. K. MURMANN (1952): J. Amer. chem. Soc. **74**, 5243.

— — (1954): J. Amer. chem. Soc. **76**, 211.

BAYLEY, S. T. (1955): Biochim. biophys. Acta **17**, 194.

BENNETT, G. M. (1941): Trans. Faraday Soc. **37**, 794.

BETHE, H. (1929): Ann. Physik (5) **3**, 133.

BIELIG, H. J., and E. BAYER (1953): Liebigs Ann. Chem. **580**, 135.

— — (1954): Experientia (Basel) **10**, 300.

BJERRUM, J. (1941): Metal ammine formation in aqueous solution. Copenhagen: P. Haase and Son.

— (1950): Chem. Rev. **46**, 381.

—, C. J. BALLHAUSEN and C. K. JØRGENSEN (1954): Acta chem. scand. **8**, 1275.

—, and E. J. NIELSEN (1948): Acta chem. scand. **2**, 297.

BJERRUM, N. (1926): Ergebn. exakt. Naturwiss. **5**, 125.

CALVIN, M., and R. H. BAILAR (1946): J. Amer. chem. Soc. **69**, 949.

CANNAN, R. K., and A. KIBRICK (1938): J. Amer. chem. Soc. **60**, 2314.

CHABEREK JR., S., and A. E. MARTELL (1952): J. Amer. chem. Soc. **74**, 6228.

COHN, M. (1954): The mechanism of enzyme action. Ed. W. E. MCELROY and B. GLASS. p. 246. Johns Hopkins Press.

—, and J. TOWNSEND (1954): Nature (Lond.) **173**, 1090.

CORNAZ, J. P., and H. DEUEL (1954): Experientia (Basel) **10**, 137.

CORWIN, A. H., and Z. REYES (1956): J. Amer. chem. Soc. **78**, 2437.

COTTON, F. A., and F. E. HARRIS (1955): J. phys. Chem. **59**, 1203.

CRAIG, P. N. (1953): Ann. N. Y. Acad. Sci. **57**, 67.

CURCHOD, J. (1956): J. Chim. phys. **53**, 17.

DATTA, S. P., and B. R. RABIN (1956): (a) Biochim. biophys. Acta **19**, 574; (b) Trans. Faraday Soc. **52**, 1123.

DAWSON, C. R., and M. F. MALLETTE (1945): Advanc. Protein Chem. **2**, 179.

DEBYE, P., and E. HÜCKEL (1923): Physik. Z. **24**, 185.

DENNEY, T. O., and C. B. MONK (1952): Trans. Faraday Soc. **47**, 992.

DERR, T. F., and W. C. VOSBURGH (1943): J. Amer. chem. Soc. **65**, 2408.

DEUEL, H., and J. SOLMS (1951): Kolloid-Z. **124**, 65.

— — (1954): Natural plant hydrocolloids. p. 62. American Chemical Society.

— — and H. ALTERMATT (1953): Vjschr. naturforsch. Ges. Zürich **98**, 49.

DE WITT, R., and J. I. WATTERS (1954): J. Amer. chem. Soc. **76**, 3810.

DIXON, M. (1953): Biochem. J. **55**, 161.

DOBBIE, H., W. O. KERMACK and H. LEES (1955): Biochem. J. **59**, 246, 257.

DOTY, P., and G. EHRLICH (1952): Ann. Rev. phys. Chem. **3**, 57.

EDELHOCH, H., E. KATCHALSKI, R. H. MAYBURY, W. L. HUGHES JR. and J. T. EDSALL (1953): J. Amer. chem. Soc. **75**, 5058.

EDSALL, J. T., G. FELSENFELD, D. S. GOODMAN and F. R. N. GURD (1954): J. Amer. chem. Soc. **76**, 3054.

FERNELIUS, W. C., and L. G. VAN UITERT (1954): Acta chem. scand. **8**, 1726.
FRANK, H. P., S. BARKIN and F. E. EIRICH (1957): J. phys. Chem. **61**, 1375.
FRANK, H. S., and M. W. EVANS (1945): J. chem. Phys. **13**, 507.
FRONAEUS, S. (1951): Acta chem. scand. **5**, 859.
— (1952): Acta chem. scand. **6**, 1200.
FUOSS, R. M., and V. F. H. CHU (1951): J. Amer. chem. Soc. **73**, 949.
GRANICK, S. (1946): Chem. Rev. **38**, 379.
GREGOR, H. P. (1956): Trans. N. Y. Acad. Sci. **18**, 667.
—, L. B. LUTTINGER and E. M. LOEBL (1955): J. phys. Chem. **59**, 34.
GURD, F. R. N. (1954): Ion transport across membranes. p. 246. Ed. H. T. Clarke, Academic Press.
—, and D. S. GOODMAN (1952): J. Amer. chem. Soc. **74**, 670.
GUTER, G. A., and G. S. HAMMOND (1956): J. Amer. chem. Soc. **78**, 5166.
HAHN, H. v., J. BÄUMLER, W. ROTH and H. ERLENMEYER (1953): Helv. chim. Acta **36**, 10.
HARES, G. B., W. C. FERNELIUS and B. E. DOUGLAS (1956): J. Amer. chem. Soc. **78**, 1816.
HARTMANN, H., and H. L. SCHLÄFER (1954): Angew. Chem. **66**, 768.
HAUROWITZ, F., and R. L. HARDIN (1953): The proteins. Vol. 2, p. 271. Ed. H. NEURATH and K. BAILEY. Academic Press.
HUGHES JR., W. L. (1947): J. Amer. chem. Soc. **69**, 1836.
— (1950): Cold Spr. Harb. Symp. quant. Biol. **14**, 79.
HUGHES, T. R., and I. M. KLOTZ (1956): J. Amer. chem. Soc. **78**, 2108.
HUIZENGA, J. R., P. F. GRIEGER and F. T. WALL (1950): J. Amer. chem. Soc. **72**, 2636, 4228.
ILSE, F. E., and H. HARTMANN (1951): Z. phys. Chem. **197**, 239.
IRVING, H., and R. J. P. WILLIAMS (1953): J. chem. Soc. 3192.
— —, D. J. FERRETT and A. E. WILLIAMS (1954): J. chem. Soc. 3494.
IZATT, R. M., C. G. HAAS, B. P. BLACK and W. C. FERNELIUS (1954): J. phys. Chem. **58**, 1133.
JONASSEN, H. B., E. R. REEVES and L. SEGAL (1955): (a) J. Amer. chem. Soc. **77**, 2667; (b) J. Amer. chem. Soc. **77**, 2748.
JØRGENSEN, C. K. (1955): Acta chem. scand. **9**, 116.
KACHMAR, J. F., and D. P. BOYER (1953): J. biol. Chem. **200**, 669.
KAGAWA, I., and K. KATSUURA (1952): J. Polymer Sci. **9**, 405.
— — (1955): J. Polymer Sci. **17**, 365.
KATCHALSKY, A., and J. GILLIS (1949): Rec. Trav. chim. Pays-Bas **68**, 879.
—, and S. LIFSON (1953): J. Polymer Sci. **11**, 409.
KERN, W. (1948): Makromol. Chem. **2**, 279.
KIMMEL, J. R., and E. L. SMITH (1954): J. biol. Chem. **207**, 515.
KING, E. L. (1949): J. Amer. chem. Soc. **71**, 319.
KLOTZ, I. M. (1953): The proteins. Vol. I, Part B, p. 727. Ed. H. NEURATH and K. BAILEY. Academic Press.
— (1954): The mechanism of enzyme action. p. 275. Ed. W. D. MCELROY and B. GLASS. Johns Hopkins Press.
—, and H. G. CURME (1948): J. Amer. chem. Soc. **70**, 939.
—, I. L. FULLER and J. M. URQUHART (1950): J. phys. Colloid Chem. **54**, 18.
—, T. A. KLOTZ and H. A. FIESS (1957): Arch. Biochem., **68**, 284.
—, and W. C. LOH-MING (1954): J. Amer. chem. Soc. **76**, 805.
—, and J. M. URQUHART (1949): J. Amer. chem. Soc. **71**, 1597.
— —, T. A. KLOTZ and J. AYERS (1955): J. Amer. chem. Soc. **77**, 1919.

Kotliar, A. M. (1954): Ph. D. Thesis. Polytechnic Inst. of Brooklyn.
—, and H. Morawetz (1955): J. Amer. chem. Soc. **77**, 3692.
Leden, I. (1941): Z. phys. Chem. **188**, 160.
— (1946): Svensk kem. T. **58**, 129.
Lehninger, A. L. (1950): Physiol. Rev. **30**, 393.
Li, N. C., O. Gawron and G. Bascuas (1954): J. Amer. chem. Soc. **76**, 225.
—, J. W. White and R. L. Yoest (1956): J. Amer. chem. Soc. **78**, 5218.
Linhart, M., and M. Weigel (1951): Z. anorg. Chem. **264**, 321; **266**, 49.
— — (1955): Z. phys. Chem. N. F. **5**, 20.
MacDougall, F. H., and S. Peterson (1947): J. phys. Chem. **51**, 1346.
Malmström, B. G. (1953): Arch. Biochem. **46**, 345.
— (1955): Arch. Biochem. **58**, 381.
Manyck, A. R., C. B. Murphy and A. E. Martell (1955): Arch. Biochem. **59**, 373.
Marcus, R. A. (1955): J. chem. Phys. **23**, 1057.
Martell, A. E., and M. Calvin (1952): The chemistry of metal chelate compounds. Prentice-Hall.
Mattock, G. (1954): Acta chem. scand. **8**, 777.
Meeks, F. R., and R. A. Marcus (1958): J. phys. Chem. (in press).
Mehl, J. W., E. Pacovska and R. J. Winzler (1949): J. biol. Chem. **177**, 13.
Mellor, D., and L. Maley (1947): Nature (Lond.) **159**, 370.
— — (1948): Nature (Lond.) **161**, 436.
Michaelis, L., and H. Pechstein (1914): Biochem. Z. **59**, 77.
Milburn, R. M. (1955): J. Amer. chem. Soc. **77**, 2064.
Monk, C. B. (1945): Trans. Faraday Soc. **47**, 285, 297.
Morawetz, H. (1955): J. Polymer Sci. **17**, 442.
—, A. M. Kotliar and H. Mark (1954): J. phys. Chem. **58**, 19.
— and E. Sammak (1957): J. phys. Chem. **61**, 1357.
Myrbäck, K. (1926): Z. physiol. Chem. **159**, 1.
— (1955): Ark. Kemi **8**, 393.
Nancollas, G. H. (1956): J. chem. Soc. 744.
O'Neill, A. N. (1955): J. Amer. chem. Soc. **77**, 2837.
Orgel, L. E. (1952): J. chem. Soc. 4756.
Osawa, F., N. Imai and I. Kagawa (1954): J. Polymer Sci. **13**, 93.
Pauling, L. (1940): The nature of the chemical bond. 2nd edition, p. 92—123. Cornell Univ. Press.
Peacock, J. M., and J. C. James (1951): J. chem. Soc. 2233.
Pecsok, R. L., and J. Sandera (1955): J. Amer. chem. Soc. **77**, 1489.
Plekhan, M. I. (1951): Zhur. Obshchei Khim. **21**, 512.
Poulsen, I., and J. Bjerrum (1955): Acta chem. scand. **9**, 1407.
Prue, J. E., and G. Schwarzenbach (1950): Helv. chim. Acta 33, 963.
Rabin, B. R. (1956): Trans. Faraday Soc. **52**, 1130.
—, and E. M. Crook (1956): Biochim. biophys. Acta **19**, 550.
Rice, S. A. (1956): J. Amer. chem. Soc. **78**, 5247.
Rosenheck, K. (1956): Personal communication.
Scatchard, G. (1949): Ann. N. Y. Acad. Sci. **51**, 660.
—, and E. S. Black (1949): J. phys. Colloid Chem. **53**, 88.
—, W. L. Hughes jr., F. R. N. Gurd and P. E. Wilcox (1954): Chemical specificity in biological interactions. p. 193. Ed. F. R. N. Gurd. Academic Press.
—, I. H. Scheinberg and S. H. Armstrong jr. (1950): J. Amer. chem. Soc. **72**, 535, 540.
Schachat, R. E., and H. Morawetz (1957): J. phys. Chem. **61**, 1177.

SCHUBERT, J. (1952): J. phys. Chem. **56**, 113.
—, and A. LINDENBAUM (1952): J. Amer. chem. Soc. **74**, 3529.
SCHWARZENBACH, G. (1950): Helv. chim. Acta **33**, 974.
— (1952): Helv. chim. Acta **35**, 2344.
— (1953): Helv. chim. Acta **36**, 23.
—, and H. ACKERMANN (1947): Helv. chim. Acta **30**, 1798.
— — (1949): Helv. chim. Acta **32**, 1682.
—, G. ANDEREGG, W. SCHNEIDER and H. SENN (1955): Helv. chim. Acta **38**, 1147.
— and W. BIEDERMANN (1948): Helv. chim. Acta **31**, 459.
—, R. GUT and G. ANDEREGG (1954): Helv. chim. Acta **37**, 937.
—, E. KAMPITSCH and R. STEINER (1945): (a) Helv. chim. Acta **28**, 828; (b) Helv. chim. Acta **28**, 1133.
— — — (1946): Helv. chim. Acta **29**, 364.
—, B. MEISSEN and H. ACKERMANN (1952): Helv. chim. Acta **35**, 2333.
—, H. SENN and G. ANDEREGG, (1957): Helv. chim. Acta **40**, 1886.
SEGAL, L., H. B. JONASSEN and R. E. REEVES (1956): J. Amer. chem. Soc. **78**, 273.
SEYMOUR, R. B., F. F. HARRIS JR. and I. BRANUM JR. (1949): Ind. Eng. Chem. **41**, 1509.
SIDDALL, T. H., and W. C. VOSBURGH (1951): J. Amer. chem. Soc. **73**, 4270.
SINGER, T. P., and E. B. KEARNEY (1954): The proteins. Vol. 2, p. 135—160. Ed. H. NEURATH and K. BAILEY. Academic Press.
SKOGSEID, A. (1948): Ph. D. Thesis. Univ. of Oslo.
SMITH, D. B., and W. H. COOK (1953): Arch. Biochem. **45**, 232.
—, A. N. O'NEILL and A. S. PERLIN (1955): Canad. J. Chem. **33**, 1352.
SMITH, E. L. (1941): Proc. Nat. Acad. Sci. **35**, 80.
— (1948): J. biol. Chem. **176**, 9.
—, and M. BERGMANN (1944): J. biol. Chem. **153**, 627.
—, N. C. DAVIS, E. ADAMS and D. H. SPACKMAN (1954): The mechanism of enzyme action. p. 291. Ed. W. D. MCELROY and B. GLASS. Johns Hopkins Press.
—, J. R. KIMMEL and D. M. BROWN (1954): J. biol. Chem. **207**, 533.
SØRENSEN, S. P. L., K. LINDERSTRØM-LANG and E. LUND (1927): J. gen. Physiol. **8**, 543.
SPANAGEL, E. W., and W. H. CAROTHERS (1935): J. Amer. chem. Soc. **57**, 929.
SPIKE, C. G., and R. W. PARRY (1953): J. Amer. chem. Soc. **75**, 2726, 3770.
STEINBERGER, R., and F. H. WESTHEIMER (1951): J. Amer. chem. Soc. **73**, 429.
STEINHARDT, J. (1941): Ann. N. Y. Acad. Sci. **41**, 287.
— (1942): J. Res. Nat. Bureau Stand. **28**, 191.
—, C. H. FUGITT and M. HARRIS (1941): J. Res. Nat. Bureau Stand. **26**, 293.
STOLL, M., and A. ROUVÉ (1935): Helv. chim. Acta **18**, 1087.
—, and G. STOLL-COMTE (1930): Helv. chim. Acta **13**, 1185.
STRAUSS, U. P., N. L. GERSHFELD and H. SPIERA (1954): J. Amer. chem. Soc. **76**, 5909.
TANFORD, C. (1950): J. Amer. chem. Soc. **72**, 441.
— (1952): J. Amer. chem. Soc. **74**, 211.
TAUBE, H. (1952): Chem. Rev. **50**, 69.
VALLEE, B. L. (1955): Advanc. Protein Chem. **10**, 317.
UITERT, L. G. VAN, and W. C. FERNELIUS (1954): J. Amer. chem. Soc. **76**, 375.
— —, and B. E. DOUGLAS (1953): J. Amer. chem. Soc. **75**, 457.
VESCIA, A. (1956): Biochem. biophys. Acta **19**, 174.
WALEY, S. G. (1953): Biochim. biophys. Acta **10**, 27.
WALL, F. T., and J. W. DRENAN (1951): J. Polymer Sci. **7**, 83.
—, and S. J. GILL (1954): J. phys. Chem. **58**, 1128.

WALL, F. T., P. F. GRIEGER, J. R. HUIZENGA and R. H. DOREMUS (1952): J. chem. Phys. **20**, 1200.
—, H. TERAYAMA and S. TECHAKUMPUCH (1956): J. Polymer Sci. **20**, 477.
WANG, J. H. (1955): J. Amer. chem. Soc. **77**, 4715.
WARNER, R. C. (1954): Trans. N. Y. Acad. Sci. **16**, 182.
—, and E. WEBER (1953): (a) J. Amer. chem. Soc. **75**, 5086; (b) J. Amer. chem. Soc. **75**, 5094.
WATSON, J. D., and F. H. C. CRICK (1953): (a) Nature (Lond.) **171**, 737; (b) Cold Spr. Harb. Symp. quant. Biol. **18**, 123.
WATTERS, J. I., and E. D. LOUGHRAN (1953): J. Amer. chem. Soc. **75**, 4819.
WEBSTER, G. C., and J. E. VARNER (1954): Arch. Biochem. **52**, 22.
WESTHEIMER, F. H., and L. I. INGRAHAM (1956): J. phys. Chem. **60**, 1668.
WILLIAMS, R. J. P. (1954): J. phys. Chem. **58**, 121.
WOLDBYE, F. (1955): Acta chem. scand. **9**, 299.
WYMAN, J. (1948): Advanc. Protein Chem. **4**, 407.
YONTE, J. H., and D. S. MARTIN JR. (1952): J. Amer. chem. Soc. **74**, 2052.

Fortschr. Hochpolym.-Forsch., Bd. 1. S. 35—74 (1958)

The Study of High Polymers by Nuclear Magnetic Resonance

By

W. P. SLICHTER

Bell Telephone Laboratories, Incorporated, Murray Hill, New Jersey, USA

With 12 Figures

Contents

1. Introduction

The familiar properties of high polymers depend greatly upon the structures and motions of the macromolecules. Often these molecular details are little understood, but their interest and importance are undeniable. Our knowledge of molecular structure and motion in polymers is based upon studies using a variety of methods, including X-ray diffraction, measurements of dielectric and dynamic mechanical absorption, and infrared spectroscopy. Each method has important limitations. For example, X-ray diffraction is ineffective in establishing the positions of hydrogen atoms, yet hydrogen is one of the most common constituents of polymer molecules. The study of molecular motion by the measurement of dielectric loss depends upon the existence of permanent electric dipoles in the substance. Yet many important polymers, such as rubber, polyethylene, polytetrafluoroethylene, and polystyrene, possess no electric dipoles if the molecules are pure. Other polymers contain electric dipoles dispersed by nonpolar segments, so that characterization of the motion of the dipoles does not necessarily describe the motion of the molecule as a whole. The polyamides and polyesters are examples of such molecules.

The rather recent development of nuclear magnetic resonance spectroscopy (NMR) offers high promise for our understanding of the structures and motions of polymer molecules. There are important limitations to this method, too, but fortunately they differ from the restrictions of the other techniques. The NMR method is often highly effective in establishing positions of hydrogen atoms, and is often able to examine motion in molecules which elude the dielectric method of measurement.

2. Magnetic Properties of Nuclei

The basic aspects of nuclear magnetism have been reviewed ably and fully by PAKE (*60*) and by ANDREW (*9*). We shall recall here only briefly some features which are close to the study of polymers.

In addition to the fundamental properties of mass and charge, the nuclei of many isotopes possess angular momentum (spin). As in the case of the electron, this angular momentum is accompanied by a magnetic moment. That is, the nuclei of these isotopes behave as if they were rotating electric charges or current loops, giving rise to magnetic moments. The magnetic moment of the nucleus is very small compared with that of the electron. The evidences of nuclear magnetism appear in rather special though important ways, as in the hyperfine structure of atomic and molecular spectra.

We shall use the proton as an example of a nucleus having a magnetic moment, since this nucleus occurs so widely in polymers. Let us imagine a hypothetical experiment in which a solitary proton is placed in the field of a laboratory magnet. Then it is a fundamental fact that the proton occupies one or the other of two energy levels in the magnetic field. In a simplified way, one may state that the magnetic moment of the proton can have either of two orientations with respect to the applied magnetic field, parallel or antiparallel. (There are other nuclei which can exhibit more than just two discrete orientations with respect to the magnetic field, but we shall not consider them, since the hydrogen and fluorine nuclei, each exhibiting only two orientations, are the only nuclei so far examined in polymers.) These two orientations differ in energy by an amount

$$\Delta E = 2\mu H_0 . \tag{1}$$

Here μ is the magnetic moment of the proton, and H_0 is the strength of the applied magnetic field. This energy ΔE, either absorbed or emitted, may be expressed through the BOHR relation in terms of a frequency, ν, and the PLANCK constant, h.

$$h\nu = 2\mu H_0 . \tag{2}$$

If the field strength is exactly 10,000 Gauss, this transition for the proton is caused by absorption or emission of energy of 42.57 mc. This is termed the resonant frequency. (The frequency is different for other nuclei. At the same value of H_0, the resonance of the fluorine nucleus, for example, occurs at 40.07 mc., and that of the phosphorus nucleus occurs at 17.24 mc.)

Because the fields of laboratory magnets are commonly in the range of 5000—10000 Gauss, resonant frequencies of nuclei lie in the megacycle portion of the electromagnetic spectrum, and so we use radio techniques of detection. The experiment is truly a form of spectroscopy, however. Returning to our hypothetical experiment in which a single proton has been placed in the magnetic field, we would find the transitions of this proton to give rise to a single spectroscopic line. Fig. 1a shows this line; the energy scale is plotted in terms of the magnetic field, with the resonance at H_0.

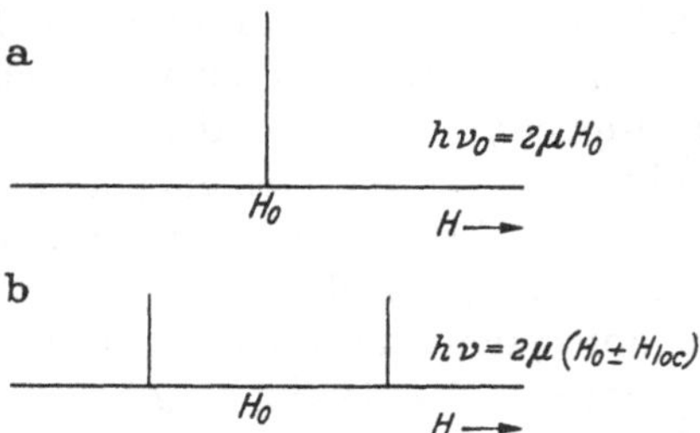

Fig. 1. Hypothetical NMR spectra: (a) single proton; (b) pair of protons

Let us now suppose that our hypothetical experiment is performed upon two protons instead of one. Then the magnetic field sensed by each proton consists of the field from the laboratory magnet supplemented by the field originating from the magnetic moment of the other proton. Equation (2) then becomes

$$h\nu = 2\mu(H_0 \pm H_{loc}) . \tag{3}$$

H_{loc}, the local field, either adds to or subtracts from the laboratory field, depending upon whether the two proton magnetic moments are parallel or opposed to one another. The size of H_{loc} depends strongly upon the internuclear distance and upon the spatial arrangement of the protons with respect to the direction of the field H_0. When H_{loc} is written out explicitly, Equation (3) becomes, according to PAKE (*59*),

$$h\,\nu = 2\mu\left[H_0 \pm \frac{3}{2}\,\mu\, r^{-3}\,(3\cos^2\theta - 1)\right] \tag{4}$$

where r is the distance between the two nuclei and θ is the angle between the internuclear vector and the laboratory magnetic field. With the two protons in our hypothetical experiment, then, the single resonance line has become split into a pair of lines equally spaced about H_0, by an amount depending upon the relative positions of the nuclei and the orientations of their magnetic moments. This splitting is shown in Fig. 1b. Typically, H_{loc} is of the order of 5 Gauss, when the nuclei are about 1 Å apart. The splitting of the resonance line offers, at least in principle, quite a precise measurement of the nuclear distance. The nuclear resonance approach is particularly attractive in the study of proton separations, because of the difficulty of locating proton coordinates by other techniques.

2.1 Absorption Spectra in Solids. Although we have dealt with a hypothetical experiment upon a solitary pair of protons, in practice there are solids in which magnetic nuclei occur in pairs, with the distance between members of each pair much smaller than the separations between pairs. The interactions between the pairs, though relatively small, are not negligible. The effect of these interactions is to cause each of the split resonances in Fig. 1b to be augmented, positively and negatively, with the result that each becomes an envelope containing a multiplicity of lines. If the solid is not a single crystal, but instead is polycrystalline, the averaging out of the angle θ over all directions tends to smear out the fine structure still further. Evidence of the pair splitting still remains, though, as is shown in Fig. 2a, a somewhat schematic view of absorption envelopes obtained with polycrystalline forms of some hydrated salts [PAKE (*59*)] and 1,2-dichloroethane [GUTOWSKY, KISTIAKOWSKY, PAKE, and PURCELL (*32*)].

Similarly, splittings of a more complicated nature have been studied for systems of *three* nuclei which may be considered to be virtually isolated in space. Again, it is possible to calculate the expected shape of the absorption envelope [ANDREW and BERSOHN (*5*)] and to make comparisons with experiment. Fig. 2b shows schematically the resonance curve for such a system, nitric acid monohydrate [RICHARDS and SMITH (*71*)].

More commonly, though, there is such a multiplicity of interactions that the absorption envelope lacks the features shown by specialized systems such as those of Fig. 2a and b. Instead, the absorption envelope is found to be featureless, as in Fig. 2c. Most polymers give rise to curves such as this, or to superpositions of such curves. With polymers, at least, these curves cannot be described by a Gaussian distribution of energies, nor indeed by any convenient analytic function.

2.2 The Effect of Motion Upon the Absorption. We have so far considered tacitly that the nuclei are motionless with respect to each other. When molecular motion occurs, the magnetic interactions between nuclei assume a time average, which as one would expect is less than the local field for a motionless system. Hence when molecular motion

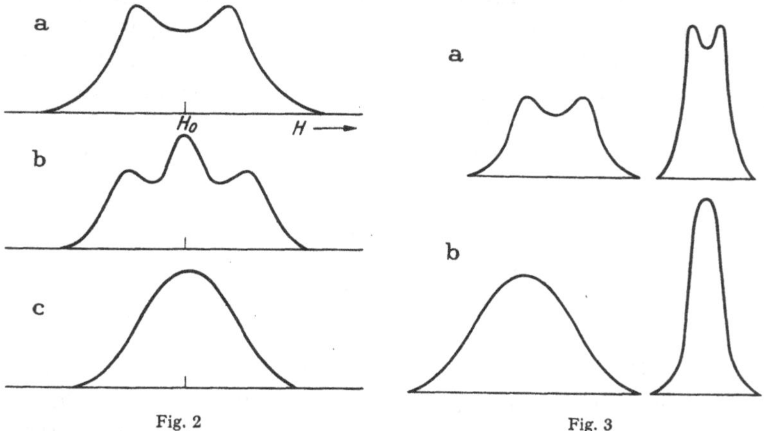

Fig. 2 Fig. 3

Fig. 2. Absorption envelopes for solids (schematic): (a) pairs of protons; (b) trios of protons; (c) complex arrays of protons

Fig. 3. Effect of motion on absorption curve, comparing rigid-lattice (left) and motion about single axis (right): (a) pairs of protons; (b) complex arrays of protons (schematic)

occurs rapidly enough, the resonance envelope becomes narrower. Such narrowing has been observed in many solids with increase in temperature. The violent motion characteristic of liquids leads to very narrow resonances, typically a thousandth or less of the width found in rigid solids.

Fig. 3a shows schematically the shape of the absorption envelope, or "line" as it is more often called, for a system of stationary nuclear pairs (Fig. 2a) and for the same system when motion occurs about an axis perpendicular to the internuclear axis [Gutowsky and Pake (*33*)]. Fig. 3b shows schematically the effect of motion upon a resonance curve of the sort depicted in Fig. 2c. The frequencies of reorientation necessary to cause a marked narrowing of the width of the envelope or "line" typically amount to 10^4 or 10^5 cycles per second [Bloembergen, Purcell, and Pound (*14*); Gutowsky and Pake (*33*)]. Such frequencies are low compared with the frequencies responsible for most thermodynamic changes. As we shall emphasize later on, one must be careful that comparisons between transitions identified by the onset of nuclear magnetic resonance (NMR) line narrowing and transitions identified by other techniques lie in the same frequency range.

2.3 The Second Moment of the Resonance Line. Although one may calculate the line shape for systems in which two or three nuclei are clustered in relatively separate groups (and for some special cases of larger groups), in general the direct calculation of the line shape is too tedious to be contemplated. More than that, the result would be rather unsatisfying, for the multiplicity of interactions between groups smears out all detailed fine structure in the line shape. For comparison between a structural model and experiment, one usually resorts instead to the *second moment* of the resonance line shape, following the treatment of VAN VLECK (*86*) and of GUTOWSKY et al. (*32*).

ΔH_2^2, the second moment expressed in terms of the magnetic field, may be defined as

$$\Delta H_2^2 = \frac{\int_{-\infty}^{\infty} f(H)\,(H - H_0)^2\,dH}{\int_{-\infty}^{\infty} f(H)\,dH}\,. \tag{5}$$

Here $f(H)$ is the line shape as a function of the magnetic field. As may be seen from Equation (5), ΔH_2^2 is the mean value of $(H - H_0)^2$, the square of the field deviation from the resonant value, with the average being taken over the symmetrical line shape, $f(H)$. The denominator appears for normalization. ΔH_2^2 can of course be evaluated from experimental curves by numerical methods, even though $f(H)$ is usually not an analytic function.

The methods of VAN VLECK's approach (*86*), circumventing the detailed calculations of the interactions, are quantum mechanical and will not be reviewed here. It suffices to note that the theory leads to an explicit expression for ΔH_2^2 in terms of the internuclear vectors and their orientations with respect to the applied magnetic field. One is therefore able to postulate a model for the structure, in which these vectors and their orientations become the parameters for the calculation, and one may test the merit of the model by comparing the calculated and measured values of ΔH_2^2. The expression for a motionless lattice is

$$\begin{aligned} \Delta H_2^2 = \frac{3}{2}\,\frac{I(I+1)}{N}\,g^2\,\beta^2 \sum_{j>k} (3\cos^2\theta_{jk} - 1)^2\,r_{jk}^{-6} \\ + \frac{1}{3}\,\frac{\beta^2}{N} \sum_{j,f} I_f(I_f + 1)\,g^2 (3\cos^2\theta_{jf} - 1)^2\,r_{jf}^{-6}\,. \end{aligned} \tag{6}$$

The symbols in Equation (6) are defined as follows:

g, I — Nuclear g-factor and spin for the nucleus at resonance.

g_f, I_f — Nuclear g-factors and spins for other ("foreign") nuclei in the sample.

β — Nuclear magneton (5.049×10^{-24} erg/Gauss; note that $g\,\beta\,I$ is the nuclear magnetic moment, μ).

r_{jk}, θ_{jk} — Magnitude of vector connecting nuclei j and k, and angle between this vector and the direction of the applied magnetic field (subscripts j, k refer to the nuclear species at resonance, and subscripts f refer to all other nuclear species present).

N — Number of nuclei at resonance which occur in the subgroup (molecule, ion, or complex) within which the line-broadening interactions are considered to take place.

By "subgroup" one means, in the case of polymers, a chain segment which contains all of the spatially non-equivalent nuclei at resonance, in the same proportion as they occur in the bulk sample. The subgroup may coincide with a chemical repeating unit, as it does with polyamides [SLICHTER (*78*)]; or it may include more than one monomer unit, as it does with polyisobutylene [POWLES (*66*)]. When more than a single equivalent position for the resonating nucleus exists in the subgroup, one must calculate the contributions for each such position, and then combine the results, with suitable weighting factors for the contributing terms, in order to get the total ΔH_2^2. The calculation in principle should be carried out over the entire sample, but because the contributions to ΔH_2^2 fall off very rapidly with increasing r, it ordinarily suffices to confine the calculation to a volume decribed by a radius of only a few Ångström units.

With crystalline powders (and hence with unoriented crystalline polymers) the terms in θ are averaged over a sphere, and Equation (6) becomes

$$\Delta H_2^2 = \frac{6}{5}\,\frac{I(I+1)}{N}\,g^2\,\beta^2 \sum_{j>k} r_{jk}^{-6} + \frac{4}{5}\,\frac{\beta^2}{N} \sum_{j,f} I_f(I_f+1)\,g^2\,r_{jf}^{-6}\,. \tag{7}$$

2.4 Calculations of Molecular Motion. We have noted that molecular motion at high enough frequency leads to narrowing of the resonance curve. In some simple systems, such as nuclear pairs or trios, it is possible to calculate in detail the line shapes corresponding to certain classes of motion. For all but the simplest systems, though, one must again make use of the second moment. The expression for the rigid lattice, Equation (6), has been extended to describe rotation about a single axis [GUTOWSKY and PAKE (*33*)], reorientation about more than one axis [POWLES and GUTOWSKY (*64*)], and also torsional oscillation [ANDREW (*6*)]. That part of ΔH_2^2 coming from within the molecule can be calculated simply enough, but the intermolecular contribution to

ΔH_2^2 is complicated by the fact that both r_{jk} and θ_{jk} vary during motion. Although the problem has been treated generally [ANDREW and EADES (*7, 8*)], the calculation is formidable. Approximate values of the intermolecular terms for ΔH_2^2 in rotating systems have been calculated for several compounds [ANDREW (*6*); ANDREW and EADES (*7*); SLICHTER (*78, 80*); SMITH (*82*)].

With polymers, rarely if ever can one say that the solid possesses predominantly a single class and extent of motion. When there is no motion, at low temperatures, calculations following Equation (6) have detailed value for comparison with experiment, to test structural models. Calculations of ΔH_2^2 for certain classes of motion are valuable not as descriptions of reality in detail but rather as gauges of the vigor of the motion.

2.5 Pulse Methods of Nuclear Resonance. So far we have dealt tacitly with steady-state experiments, in which the nuclei are continuously exposed to radiofrequency energy at the resonant value. Most NMR studies on polymers to date have been made under steady-state conditions. A separate class of studies involves the application of a brief pulse of radiofrequency energy, following which the transient response is studied [HAHN (*37*); CARR and PURCELL (*20*)]. Information on molecular motion is often obtainable from such experiments. Pulse ("spin-echo") studies on polymers have thus far been reported only fragmentarily [NOLLE (*55*); LOWE, BROWN and NORBERG (*45*); PAKE (*61*); WILSON and PAKE (*91*); GUTOWSKY, SAIKA, TAKEDA and WOESSNER (*36*)], and will not be reviewed here. They are mentioned in passing because of the potential importance of this approach to the study of polymers.

3. Experimental Methods

Steady-state experiments on solids fall into two classes: (a) nuclear resonance absorption; (b) nuclear induction. Both methods yield the same information. In exploring the resonance line, we may vary either the radio frequency or the magnetic field. Usually it is more convenient to keep the frequency constant and to vary the magnetic field over a suitable range about the resonant value.

3.1 Nuclear Resonance Absorption. In a typical apparatus for nuclear resonance absorption, Fig. 4 [POUND and KNIGHT (*63*); WATKINS and POUND (*87*); GUTOWSKY, MEYER and MCCLURE (*34*)], the sample is contained in a small coil situated in the magnetic field. The coil is part of the tuned circuit of a radiofrequency oscillator. The magnetic field is changed slowly and linearly in time. The field is also modulated at low audio frequency and with an amplitude which is small compared with the resonance line width. When the magnetic field passes through the resonant value, the sample absorbs radio energy, thus changing the amplitude of the r. f. oscillation. Since the magnetic field is continuously being modulated at audio frequency, this change in the amplitude of oscillation contains an audio component which is detected and then amplified. The use of narrow-band amplification at this audio frequency, synchronized or "locked-in" with the phase of the modulation, greatly

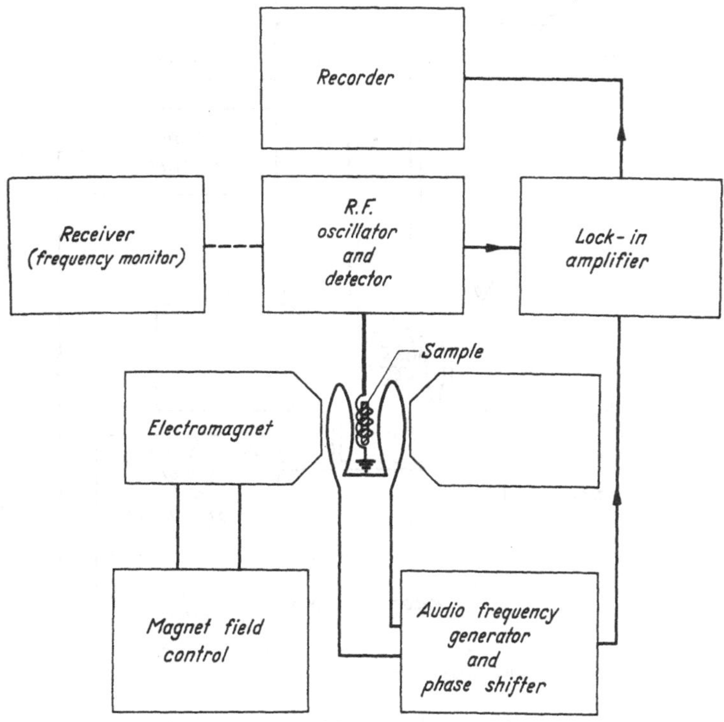

Fig. 4. Block diagram of typical nuclear resonance absorption apparatus

enhances the ratio of signal to noise. The output of the amplification is plotted on a recording potentiometer. As a consequence of this system of modulation and detection, it is not the absorption curve itself which is recorded, but rather the derivative of the absorption with respect to the magnetic field. In most instances the interpretation of the data can be made directly in terms of the derivative curve, without integration to obtain the absorption curve.

3.2 Nuclear Induction. The method of nuclear induction [BLOCH, HANSEN and PACKARD (*13*)] involves the use of two coils to contain the sample (Fig. 5). Radiofrequency energy is supplied by the transmitting coil, producing transitions in the orientations of the nuclear moments.

The receiving coil, with its axis perpendicular to the axis of the transmitting coil and also to the direction of the steady magnetic field, picks up energy emitted by these nuclear transitions. As with the absorption method, one commonly modulates the magnetic field at audio frequency and small amplitude, and then amplifies the detected radio signal by means of a phase-sensitive, narrowband circuit.

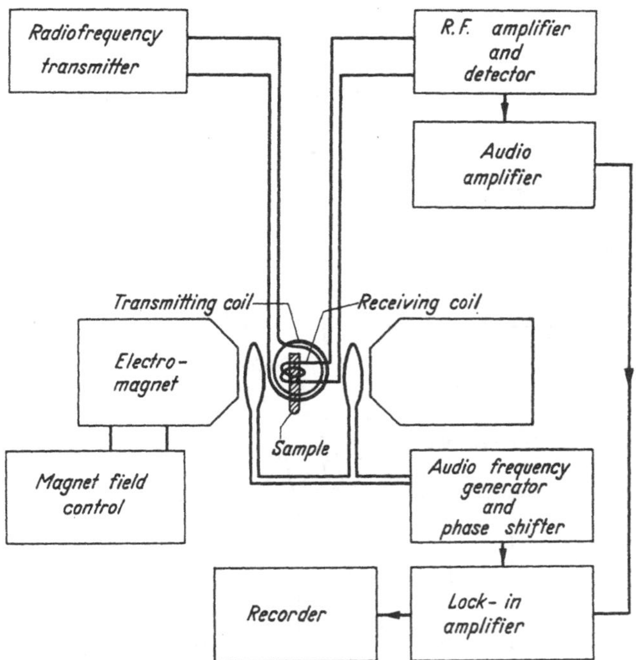

Fig. 5. Block diagram of typical nuclear induction apparatus

Other circuits, such as those involving radiofrequency bridges [BLOEMBERGEN et al. (*14*); ANDERSON (*4*)], are not greatly different from those just mentioned. For the study of polymers there is no special advantage to any one method. The nuclear induction method is somewhat less convenient than the methods involving a single coil when the experiment requires a cryostat.

4. Early Studies on Polymers

When the value of the nuclear magnetic resonance method for the study of transitions in solids became evident, it was natural to extend these studies to polymers. ALPERT (*2, 3*) noted that the resonance line for natural rubber, which was very narrow at room temperature, became somewhat broader at lower temperature. He also noted some broadening of the resonance line when the rubber was loaded with carbon black, but he detected no effect of vulcanization upon line width. Similar results were reported by MROWCA, HOLROYD and GUTH (*51*). The temperature variation of the line width in polyethylene has been described by NEWMAN (*52*). This polymer, presumably a highly-branched polyethylene, showed marked narrowing of the resonance line over a 20° span somewhat below room temperature.

5. Elastomers

So far, nuclear resonance studies on elastomers have largely been qualitative, and indeed in several instances have led to unsound conclusions. There is every reason to expect, though, that systematic NMR studies on well-characterized compounds will greatly help our understanding of the motion of rubbery molecules and the effects of vulcanization and reinforcement.

5.1 Vulcanization. Vigorous molecular motion, approaching that expected for a viscous liquid, occurs in unvulcanized natural rubber (Hevea) and GR-S (a styrene-butadiene copolymer, commonly containing about 25% styrene) at room temperature, as shown by the extreme narrowness of the resonance curves [HOLROYD, CODRINGTON, MROWCA and GUTH (*38*)]. Only a small increase in the room-temperature line width is noted after vulcanization. This finding is not too surprising, for ordinarily the sulfur content of cured rubber amounts to only a few percent. That is, ordinarily there is only one sulfur atom for about ten isoprene units. Although this ratio is enough to change the mechanical properties of rubber and GR-S, the ability of the chains to move in rotation and Brownian translation is affected rather little. In view of the short-range nature of the interactions between nuclear magnetic moments (see above), it is not surprising that constraints occurring only every ten monomer units along chains which otherwise are actively moving should have rather little effect upon the NMR experiment.

There are other elastomers, though, in which the chain motion is evidently less active than in rubber. With Hycar OS-10, a 1:1 copolymer of butadiene and styrene, the room-temperature line width is several times as great as in rubber or GR-S (HOLROYD et al. (*38*)]. Moreover, vulcanization of Hycar OS-10 leads to a marked increase in line width at all temperatures compared with the line width in the uncured elastomer. These qualitative differences between elastomers point to the importance of making detailed studies which will separate the effects of chain stiffness and cross-linking.

GUTOWSKY and MEYER (*35*) have studied the line width in natural rubber over a broad temperature range, in samples cured at 280° F for 30, 60 and 90 min. The sulfur content is not stated, nor is there mention of what accelerator was used, if any. The samples must therefore be considered ambiguous. GUTOWSKY and MEYER find two regions of change in the nuclear resonance line width, one occurring at about 220—230° K and the other in the range 143—173° K. The line width in this latter region differs in the three samples, being greatest in the sample cured for 90 min. The authors note that this transition range, 143—173° K, corresponds to transition temperatures found in NMR studies on some methyl-substituted methanes [POWLES and GUTOWSKY (*64*)], and they conclude by analogy that the line narrowing seen at these temperatures in rubber comes from the onset of motion of the methyl groups about the C_3 symmetry axis. The same sort of conclusion is reached by BANAS, MROWCA and GUTH (*11*). The line narrowing found by GUTOWSKY and MEYER at about 225° K bears little relation to the cure time. Since this narrowing occurs at a higher temperature than the other transition, they conclude that it relates to a more restricted motion than methyl rotation, and they infer that this motion is in the chain segments.

Since it is the low-temperature narrowing, ascribed to methyl rotation, which varies with the time of cure, GUTOWSKY and MEYER suggest that cross-linking is not the primary mechanism of vulcanization, but that instead sulfur-containing rings within single molecules affect chain stiffness and also rotation of methyl groups. It is not fitting to review here the still-puzzling question of the nature of vulcanization. As for these nuclear resonance results, though, it should be remarked that not only do they deal with rather ambiguous vulcanizates, but that even a well-defined compound cannot be termed typical of a process which varies in such a complex fashion with composition and conditions (see below). Indeed, it may well be that the effects seen in these NMR studies are not a result of the curing itself, but rather of some simultaneous but unrelated change in the hydrocarbon.

It should also be pointed out that although line narrowing occurs in rubber at about the same temperature as in simple methyl-substituted compounds, it is risky to infer that the transitions in both kinds of compounds come from the same class of motion. The methyl-substituted compounds studied by POWLES and GUTOWSKY are wholly crystalline at the temperatures of interest. Rubber is only partially crystalline at even these low temperatures. Transitions in the physical properties of rubber occur over a broad temperature range, and doubtless represent a broad spectrum of molecular motion. Indeed, there are methyl-containing polymers (see below) which show rotation of the methyl groups at temperatures much lower than that for the transition found in rubber. These remarks on the hazards of comparing NMR transitions are of course not limited to the case at hand.

Subsequently, additional studies on rubber were made by GUTOWSKY, SAIKA, TAKEDA and WOESSNER (*36*). In these studies the compositions and properties of the vulcanizates were more explicitly established than in the previous studies. A mild acceleration of the curing was used (mercaptobenzothiazole as an accelerator with zinc oxide as an activator, the accelerator content being rather less than that used commercially). Three ranges of initial concentration of sulfur were used. The lowest corresponded roughly to those commonly used for soft rubber gum stocks, while the other two corresponded to semi-hard rubbers. On the basis of the data reported for physical properties and combined sulfur, it appears that the vulcanizates for the most part fell short of optimum cure, though the studies included some samples representative of overcure and accompanying degradation.

As we have seen, GUTOWSKY and MEYER attribute the room-temperature changes in line width to motions of the main chains, and postulate that such motions are more likely to be constrained by cross-linking than by intrachain cyclization during vulcanization. The room-temperature absorption curves obtained by GUTOWSKY, SAIKA et al. are of course quite narrow (10 to 20 milligauss). Still, these authors find discernible differences in line width for the several mixes. They note that for a given percentage of combined sulfur the absorption curves increase in width with increase in fraction of raw sulfur. They adduce that an increase in raw sulfur concentration favors cross-linking over cyclization (the latter process having been invoked by GUTOWSKY and MEYER, as we have seen, as the chief reaction in what evidently were light vulcanizations). It is well known in rubber technology, though, that combined sulfur is by no means a precise description of the state of the vulcanizate: depending upon the composition of the mix and the conditions of cure, one may arrive at compounds which display widely different physical properties but which contain the same amount of combined sulfur. Thus, the samples used by GUTOWSKY et al. display, at the same combined sulfur, both undercure and considerable overcure. Extremes in curing undoubtedly entail not only marked differences in vulcanization but also important changes in the hydrocarbon itself, from effects such as degradation.

The studies of GUTOWSKY et al. also include measurements of line width over the range 80 to 300° K. They find that the general features of the graphs of line width versus temperature resemble those described earlier by GUTOWSKY and MEYER. They note, though, that the pronounced changes ascribed earlier to the onset of methyl group rotation are absent except in samples having low percentages of combined sulfur. In general the resonance narrows at a somewhat higher temperature the greater the raw sulfur content and the greater the concentration of combined sulfur. The hazard of interpreting these general findings in terms of postulated mechanisms of vulcanization has already been mentioned.

Although the nuclear resonance results of GUTOWSKY et al. appear to have been by no means definitive in clarifying mechanisms and structures in vulcanization, still they are valuable for outlining some of the approaches for future study. The process of vulcanization ought to be looked upon as a general structural change, rather than as a restricted chemical action [FLORY (*26*)]. Clearly there is need for a thorough investigation, with nuclear resonance methods applied to samples representing wide varieties of composition, accelerator type, and conditions of cure.

5.2 Some Effects of Chain Constitution. Hycar OR-15, a copolymer of acrylonitrile and butadiene in the ratio 2:3 by weight, has been compared [HOLROYD et al. (*38*)] with the styrene-butadiene copolymer, Hycar OS-10. The transition in line width occurs at a somewhat higher temperature in the former than in the latter; this difference is ascribed by the authors to interactions between the nitrile groups in the Hycar OR-15, which constrain chain motion in this elastomer compared with that in the styrene copolymer. The interpretation is probably more complicated than this, however, for one would expect that the bulkiness of the phenyl groups in Hycar OS-10 would likewise hamper chain motion to an important degree. The differences between observed line width transitions for these two polymers doubtless include several effects, such as motion of the phenyl groups about the twofold axis. Clearly it is futile to try to interpret these constraints unambiguously on the basis of a comparison between two distinctly different compounds. It would surely be instructive, though, to employ NMR for a study of chain motions in a series of copolymers made up of a number of different ratios of the same monomers.

Chain motion has already been compared qualitatively in two elastomers copolymerized from the same monomers. HONNOLD, MCCAFFREY and MROWCA (*39*) have examined the line width as a function of temperature in Paracril-26 and Paracril-35, copolymers of acrylonitrile and butadiene containing 26 and 35% of acrylonitrile by weight, respectively. The onset of molecular motion, as shown by the narrowing of the resonance line, occurs at a somewhat higher temperature in the elastomer which is richer in acrylonitrile. The authors ascribe the difference in the transition temperature to a difference in the coupling between the polar nitrile groups in the two elastomers; they suggest that the greater dipolar content of Paracril-35 compared with that in Paracril-26 increases the constraints to molecular motion.

GUTOWSKY, SAIKA, TAKEDA and WOESSNER (*36*) have compared line width as a function of temperature for uncured Hevea rubber, comparing a sample which had been quenched in liquid nitrogen with one which was annealed at 259° K. The latter shows the same behavior in line width as the original material before thermal treatment, but the quenched sample exhibits a narrower line at all temperatures below 255° K, and a somewhat lower transition temperature for line width. The differences are ascribed to the existence of more rapid chain motions, at a given temperature, in the amorphous portions than in the crystalline.

These authors have also examined balata and synthetic poly-cis-isoprene. They report that balata unaccountably undergoes a line-width transition at lower temperature than does uncured Hevea, even though the physical properties of balata (at room temperature) might lead one to expect otherwise. Evidently there are also differences, still to be understood, in the motional line narrowing of Hevea and synthetic poly-cis-isoprene.

5.3 Reinforcement. It is well known that the addition of carbon black to Hevea and to synthetic rubbers markedly affects physical properties, and that the reinforcement depends in complicated ways upon the identity of the carbon black and the nature of the polymer. It is not appropriate to review the theories here, except to note that generally reinforcement is supposed to involve some sort of cohesion, chemical or physical, between the rubber molecule and the filler particle. If this cohesion imposes much constraint upon the motion of the polymer molecules, one would expect to find the resonance line broadened compared with its width for the unloaded polymer. HONNOLD et al. (*39*) have examined the room-temperature line width for Hevea, GR-S, butyl rubber (GR-I), and Hycar OS-10, loaded with different carbon blacks and with different proportions of black (up to $^1/_3$ by weight). In particular, these authors have examined butyl rubber with three widely different blacks, over a temperature range of 220 to 340° K. They report no appreciable increase in line width after addition of the black. Hence carbon loading, even though important to the physical properties of the elastomer, causes rather minor changes in the molecular motion seen by NMR. The constraints to motion seem to be localized near the points of attachment. HONNOLD et al. (*39*) also report some decrease in the intensity of the absorption with loading. There is doubt as to the correctness of their conclusions. Speculating as to whether this intensity decrease is to be attributed to a few strong (chemical) bonds between filler and elastomer or a number of weak (physical) bonds, the authors favor the latter view, citing the fact that the effect of reinforcement is decidedly less than the effect of vulcanization upon the NMR curves. The over-all conclusion may indeed be right, but the deduction from the

NMR data seems wrong. Marked differences in intensity between two different absorption curves result because (a) there is much more motion in one case than in the other; (b) there are more resonating nuclei in one case than in the other. The *cohesive* strength of the attachment between the carbon particle and the rubber molecule is not at all the same as the constraint to chain rotation and flexing. It seems likely that the observed decrease in the intensity of NMR curves upon carbon loading arises in large part from the reduction in the number of protons per unit volume. So far the nuclear resonance information on rubber reinforcement is sparse and purely qualitative, but more detailed studies ought to add to our understanding of the reinforcement process.

5.4 Some Effects of Stresses and Solvents. Elongation of rubber alters the line width markedly. OSHIMA and KUSUMOTO (*58*) have studied a vulcanized sample of rubber over a wide temperature range, with the material unstretched and then with it elongated 500%. They find that the resonance curve narrows at a higher temperature in the stretched sample than in the unstretched, and they take this behavior to show that orientation hinders motion of chain segments. They also find that the low-temperature line width and second moment are greater in the stretched sample than in the unstretched. Apparently these changes come about because orientation causes the chains to pack more closely than in the unstretched state.

On the other hand, the use of swelling agents greatly increases the freedom of polymer chain motion. HOLROYD et al. (*38*) have studied the resonance line width in Hycar OS-10 over a broad range of temperatures and with the addition of varying amounts of benzene as a swelling agent. When 5 parts of benzene are added to 100 parts of polymer, the NMR line width at 240° K is reduced from 8 Gauss, for the pure elastomer, to only 1.8 Gauss. Similar reductions occur at other temperatures. Furthermore, the line narrowing occurs at a lower temperature when more solvent is used. The authors attribute this decrease in line width with solvent addition to the swelling process itself, i. e., to increases in the effective distance between chains. This view is untenable; for reasonable estimates as to the intermolecular contribution to the line broadening, following VAN VLECK's treatment (*86*), lead to the unacceptable conclusion that the chain-chain distance at 240° K becomes increased by two-thirds with the addition of only 5% benzene to the solid polymer. The correct interpretation is that addition of solvent enhances the rotational and limited translational motion of the polymer segments.

In any case, the ability of solvent molecules to enhance the motion of polymer molecules is plain. The picture is supported by the detailed studies of POWLES (*66*) upon polyisobutylene swollen with benzene (see below); here a marked decrease occurs in the line width and in the second moment (Fig. 6) at all temperatures, upon addition of one molecule of solvent to 4.88 monomer units of the elastomer.

5.5 Some Studies of Chain Configuration and Motion. The study by POWLES (*66*) of the molecular structure and motion of polyisobutylene (PIB) is more quantitative than any other NMR study of an elastomer to date. X-ray diffraction studies [FULLER, FROSCH and PAPÉ (*29*); BUNN (*15, 17*); LIQUORI (*44*)] show that PIB is amorphous at room-temperature in the absence of stress, but that it becomes highly crystalline under extreme elongation. The diffraction studies all show that the

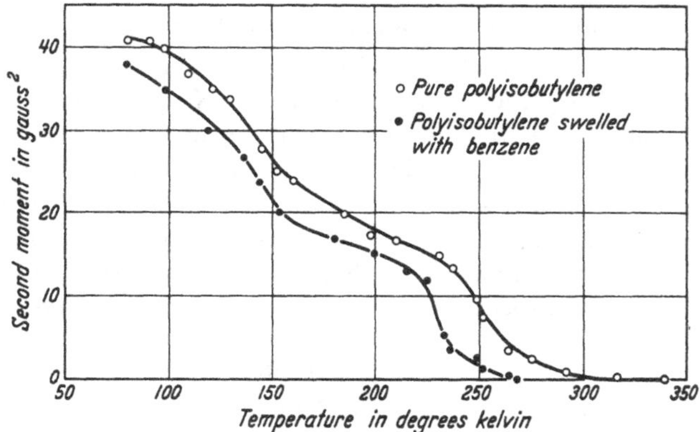

Fig. 6. Variation of second moment with temperature for polyisobutylene, pure polymer and polymer swelled with benzene [POWLES (*66*)]

chains have helical configurations in the stressed state, though there is some disagreement as to the atomic coordinates. The measured value of the NMR second moment, ΔH_2^2, at 77° K for unoriented PIB is unusually large, and is explained by POWLES as arising from close interactions between the methyl groups on alternate carbon atoms of the chain. Comparisons between the measured values of ΔH_2^2 and values calculated for some simple models confirm that the chain cannot be straight, to any important degree, for such a configuration leads to values of ΔH_2^2 which are much too high. It is not possible to calculate ΔH_2^2 accurately for the unoriented state, since there is no detailed information as to the configurations. POWLES has calculated ΔH_2^2 for the several structures which come from X-ray diffraction studies on stretched PIB. It is interesting that these alternative values for oriented PIB, which do not differ greatly from one another, are quite close to the value of ΔH_2^2 measured in the amorphous polymer at 77° K. This finding suggests that successive segments of a chain in the amorphous polymer are disposed with respect to one another in a manner which approximates the configuration of a molecule in the crystalline state.

POWLES notes that the marked decrease in line width and second moment which occur in some methyl-containing compounds at about 125° K is comparatively minor in PIB. As we have already noted, this transition has been attributed in other compounds to the onset of rotation of the methyl group about the C_3 axis. In PIB, though, the resonance curve changes continuously over a broad temperature range (Fig. 6). The absence of a pronounced transition is taken by POWLES to show that the methyl groups are strongly interlocked. This finding agrees with the evidence for steric hindrance between methyl groups in PIB which came forth earlier from studies of the heat of polymerization [EVANS and POLANYI (*24*)] and the thermal conductivity [REHNER (*69*)].

On the other hand, ROCHOW and LE CLAIR (*73*) report that in silicone rubber, $[(CH_3)_2SiO]_x$, the value of ΔH_2^2 at 77° K agrees with the figure expected if all the methyl groups are rotating about their C_3 axes. Furthermore, the resonance line narrows abruptly at 164° K with increasing temperature, some 60° below the analogous narrowing in Hevea rubber. The authors cite the difference in the transition temperature for silicone rubber from that in Hevea as explaining the difference in the low-temperature utility of these two elastomers.

6. Glassy Polymers

The glassy polymers have received rather little attention in NMR studies thus far. Our notions of chain structure from the sources such as X-ray diffraction are not clear-cut, and so it is hard to make quantitative tests of structural models by nuclear resonance. Nevertheless, it is likely that NMR methods will soon prove helpful for detailed studies such as investigation of stress relaxation or of plasticizer action in these polymers.

6.1 Polystyrene. The proton resonance absorption of polystyrene has been studied by HOLROYD et al. (*38*). They note that the temperature at which marked narrowing occurs is a few degrees higher in a fractionated sample of high molecular weight (1,500,000) than in a sample of lower molecular weight (447,000). But even the latter fraction possesses a reasonably high molecular weight. When one recalls that nuclear magnetic interactions fall off rapidly with distance, it seems strange that these differences in average molecular size between the two fractions should have a detectable effect on the resonance curves. It seems likely that the differences seen by HOLROYD et al. come from details of the molecular weight distributions, rather than from the shift in the average molecular weight. HOLROYD et al. (*38*) find that the line width transition occurs over a shorter temperature range in the two fractionated samples than unfractionated polystyrene. Clearly more work needs to be done to distinguish between the effects of average molecular weight and the distribution of molecular weight.

In all the samples studied by HOLROYD et al., the line narrowing occurs at higher temperatures, about 390 to 400° K, than in the familiar elastomers or in many crystalline polymers (see below). Presumably this elevated transition temperature reflects the constraint to chain motion imposed by the bulkiness of the phenyl group. A much less pronounced narrowing is seen to begin at a lower temperature, about 350° K. Similar results are reported by ODAJIMA, SOHMA and KOIKE (*57*). These authors associate this low-temperature change with the glass transition, which has been identified at 351° K in the dilatometric studies of PATNODE and SCHEIBER (*62*), but it seems likely that in the samples studied by NMR the glass temperature was about 375° K (see below).

ODAJIMA, SOHMA and KOIKE have calculated the second moment of polystyrene, but in order to make the calculation tractable they have assumed the molecule to be fully stretched and in planar zig-zag form. As the authors themselves point out, such a model can hardly be realistic, and so it is not surprising that poor agreement is found here between calculation and experiment. Indeed, very little is known of the details of the molecular structure and configuration of this polymer in the glassy state [SCHILDKNECHT (*76*)]. The difficulties with polystyrene illustrate the point that the NMR method is poorly suited to establishing structures in the absence of rather explicit information from other sources. No doubt the method will prove valuable, though, in studies of *isotactic* polystyrene, on which detailed work is now being done in a number of laboratories by other methods.

6.2 Polymethyl Methacrylate (PMMA). The proton resonance in PMMA has been studied by HOLROYD et al. (*38*); by ODAJIMA, SOHMA and KOIKE (*56, 57*); and by POWLES (67). All of these studies show a marked narrowing of the resonance line shortly above 340° K, but there are distinct differences between the line widths reported by the three groups of investigators. POWLES suggests that these disparities stem from differences between the samples. If so, they probably relate to differences in the molecular weight and particularly the distribution of molecular weight.

POWLES has measured the second moment of the resonance over a broad temperature range (Fig. 7). He finds a decrease in ΔH_2^2 with rising temperature in the interval 130 to 200° K, the temperature region associated with the onset of methyl group rotation in some methyl-containing compounds (see above). POWLES has calculated ΔH_2^2 for PMMA, presuming the molecules to be motionless. The scheme of the calculation is that the values of ΔH_2^2 for isolated CH_2 and CH_3 groups

[GUTOWSKY and PAKE (33); ANDREW and BERSOHN (5)] are weighted according to the proportion of these groups in the PMMA molecule, and are augmented by a figure estimated to represent the interactions between proton groups within the molecule and between molecules. POWLES is forced to this admittedly inexact calculation because of the lack of definite models for the chain configuration, yet the result agrees satisfactorily with the value measured at 77° K.

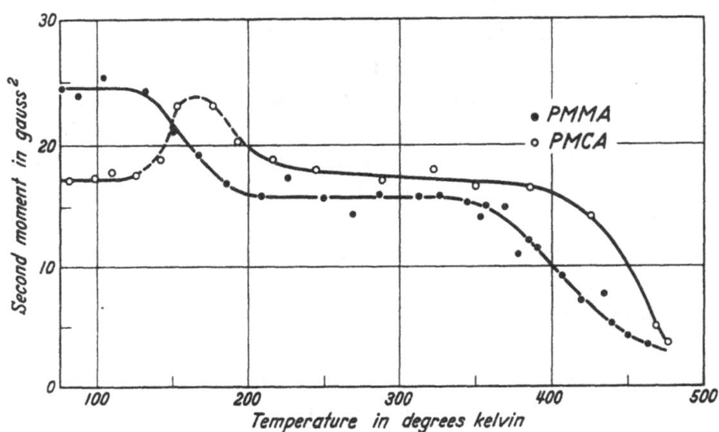

Fig. 7. Variation of second moment with temperature for polymethyl methacrylate and polymethyl α-chloracrylate [POWLES (67)]

POWLES has also calculated ΔH_2^2 for two examples of chain motion, one in which half of the methyl groups rotate rapidly about their C_3 axes and the other in which all the methyl groups rotate. He finds that the measured value of ΔH_2^2 from 200 to 340° K (Fig. 7) corresponds closely to the value he predicts for motion of half of the methyl groups. From this result he postulates that the two types of methyl groups in PMMA, those attached to the main chain and those attached to the branches, have different but essentially uniform environments throughout the solid, even though there is no structural regularity of the sort which would be seen by X-ray diffraction.

POWLES suggests that this change in ΔH_2^2 (Fig. 7) comes about because one class of methyl groups, perhaps those attached to the side chains, undergoes rotation while the other class of methyl groups is effectively stationary. He has sought to distinguish between the motions of the two types of methyl groups by examining a closely related polymer, polymethyl α-chloroacrylate (PMCA), in which chlorine atoms replace the methyl groups attached to the main chain in PMMA. The results of this part of POWLES' study are ambiguous (see below), but the matter deserves further work.

One would have high confidence in a model of the molecular structure if one could use such a model to predict the actual line shape and could then confirm the result by experiment. We have already noted that the difficulties in calculating the line shape led to the development of the theory of the second moment. The shortcomings of a simplified calculation of the line shape are demonstrated by POWLES' attempt at calculating the line shape for PMMA. He evaluates the line shape from a superposition of the theoretical shapes for isolated CH_2 and CH_3 groups [PAKE (*59*); ANDREW and BERSOHN (*5*)], with these curves broadened by a Gaussian term to account for interactions between groups. The comparisons between theory and experiment are rather disappointing. As POWLES himself points out, the use of the Gaussian broadening term is doubtless incorrect, for the interactions between groups can scarcely be the same as for a random arrangement of protons.

6.3 Polymethyl α-chloroacrylate (PMCA). Fig. 7 also shows POWLES' results for PMCA. The increase in ΔH_2^2 with decrease in temperature, which occurs down to about 165° K, is qualitatively what one would expect. The decrease in ΔH_2^2 below this temperature is puzzling, for it would seem to say that molecular motion increases as the temperature is lowered. POWLES feels that the effect is independent of the history of the sample. He suggests that it may stem from a phase transition occurring over a broad temperature range, though he points out that there is no evidence of such a transition from other sources. It should also be noted that only rarely do transitions lead to an increase in line width and second moment as temperature is increased [TAKEDA and GUTOWSKY (*83*)]. Alternatively, POWLES suggests that the behavior may be a special case of "saturation". The phenomenon of saturation, which occurs in NMR experiments when the amount of radiofrequency energy supplied to the nuclei is enough to alter the steady-state population of nuclei in the different levels, is too detailed to be treated in this review; the reader is referred to more specialized reports [BLOEMBERGEN et al. (*14*); ANDREW (*9*)]. As POWLES himself points out, though, saturation ordinarily results in a broadening of the resonance curve, not a narrowing. Hence if saturation accounts for the low-temperature behavior in PMCA, the effect is unique.

6.4 The Glass Temperature. As many workers have shown, there exists for every amorphous polymer a rather narrow temperature region within which the substance changes from a brittle, glassy substance below this temperature zone, to a viscous or rubbery material above the zone. (There is some evidence of such a change in some semicrystalline polymers, too; the effect has been ascribed to the amorphous regions of these polymers.) This *glass temperature*, T_g, marks the approximate temperature for the change of a variety of properties, including hardness, refractive index, specific volume, permeability to small molecules, dielectric constant, and dynamic mechanical modulus. It does not correspond to a phase transition (hence some authors refer to it as the second-order transition temperature); indeed, values of T_g depend on the method of study, the time scale of the experiment, the history of the sample, and of course the nature of the polymer.

Despite the fact that one can state only approximate values for T_g, the transformations marked by this temperature undoubtedly describe characteristic changes in the motions of portions of the molecules. Furthermore, comparisons of T_g for different polymers help describe the relative degree of constraint to chain motion. It is natural that NMR should be added to the other methods of study of T_g. For example, the marked narrowing of the NMR absorption of PMMA [HOLROYD et al. (*38*); ODAJIMA et al. (*56, 57*); POWLES (*67*)] in the interval 350 to 380° K corresponds fairly closely in temperature to glass transitions found by other methods [WÜRSTLIN (*92*); ROGERS and MANDELKERN (*74*)]. Also, as POWLES (*67*) points out, the NMR transition occurs over the same temperature interval as that in which dielectric loss maxima have been found at frequencies of 10^4—10^5 c. p. s. [DEUTSCH, HOFF and REDDISH (*23*)]. The dielectric results are of course sensitive to the motion of the ester group, since it contains the electric dipole. As we have noted, these frequencies also correspond typically to the frequencies responsible for NMR line narrowing. The body of evidence would seem to say, then, that the glass transition in PMMA corresponds to motion of the entire ester group.

Similarly, the NMR transition observed in polystyrene in the interval 380—390° K (HOLROYD et al. (*38*); ODAJIMA et al. (*56, 57*) corresponds closely in temperature to the glass transition seen in dilatometric studies [FOX and FLORY (*27*)]. In their NMR studies, ODAJIMA et al. (*57*) associate a lesser NMR transition near 350° K with the glass transition (see above). However, their comparison is with the early work of PATNODE and SCHEIBER (*62*), which displayed a lower value of glass temperature, in a polymer which presumably was low in molecular weight. In polyvinyl chloride, moreover, the NMR transition at about 370° K [ODAJIMA et al. (*56, 57*)] occurs near the transformation observed by refractometric studies [WILEY and BRAUER (*88*); KNAPPE and SCHULZ (*41*)] and by specific heat measurements [ALFORD and DOLE (*1*)]. Likewise, there is a fairly close agreement between the transition temperatures in butadiene – styrene copolymers studied by NMR [HOLROYD et al. (*38*); HONNOLD et al. (*39*)] and by dynamical mechanical measurements [NIELSEN, BUCHDAHL and CLAVER (*53*)].

The coincidence of transformation temperature as seen by NMR and by some other method must not be taken as evidence that the same molecular mechanism prevails in each case. For example, one must be skeptical that any close relation exists between the molecular motions responsible for the mechanical loss maxima occurring in polystyrene at a temperature of about 380° K and less than one c. p. s. [SCHMIEDER and WOLF (*77*)], and the motion responsible for the NMR line narrowing in the same temperature region but arising from motions of the order of 10^5 c. p. s. [HOLROYD et al. (*38*)]. Vagueness in the correlation between NMR transitions and glassy transformations seen by other means occurs elsewhere. For example, there is no marked transition in the NMR absorption for polyisobutylene [POWLES (*66*)] which corresponds to the glassy change seen in dilatometric studies [UEBERREITER (*85*)]. Similarly, the NMR narrowing found in polyvinyl acetate at about 350° K [TANAKA, YAMAGATA, and KITTAKA (*84*)] lies some 60° above the glassy changes reported elsewhere for this polymer [JENCKEL (*40*); WILEY and BRAUER (*89*)].

Perhaps some of the instances of agreement between NMR transitions and conventional glassy transformations are fortuitous. By the same token, the disparities between NMR studies and other results may be ascribed at least in part to differences between the time scales of the experiments and between the detailed composition of the substances. Clearly, though, the NMR method will play an increasingly important part in future studies of the glassy transformation.

7. Crystalline Polymers

Although no synthetic polymer is wholly crystalline, the crystalline content of many polymers is high enough that one may be rather definite about the molecular configuration. X-ray diffraction studies have described the crystalline structures of a number of polymers in great detail. NMR studies promise to add to the X-ray information, particularly in fixing proton coordinates, and have already been valuable in describing molecular motion. NMR studies have been valuable in describing molecular motion in some n-paraffins of comparatively low molecular weight [ANDREW (*6*)] and in cetyl alcohol [KOJIMA and OGAWA (*42*)].

7.1 Polyamides. A number of linear polyamides have been studied over a broad temperature range by SLICHTER (*78*). He finds that marked narrowing of the resonance line occurs at higher temperatures (typically 400 to 450° K) than in many polymers. Evidently this elevated temperature of transition comes about because of constraints to chain rotation imposed by the well-known coupling between amide groups in neighboring chains. Thus, in a series of polyamides crystallizing in the same habit, the transition occurs at a lower temperature as the proportion of amide groups is reduced, i. e., as the paraffinic segments are made longer (polydecamethylene octadecanediamide compared to polyhexamethylene adipamide). Similarly, when the association between the amide groups of neighboring chains is partially disordered by copolymerization [BAKER and FULLER (*10*)], the transition temperature is lowered. These effects of the coupling between amide groups upon the motion of the paraffinic segments agree with what one would expect, in light of the well-known effects of the coupling upon melting points, elastic moduli, and solubility in these polymers.

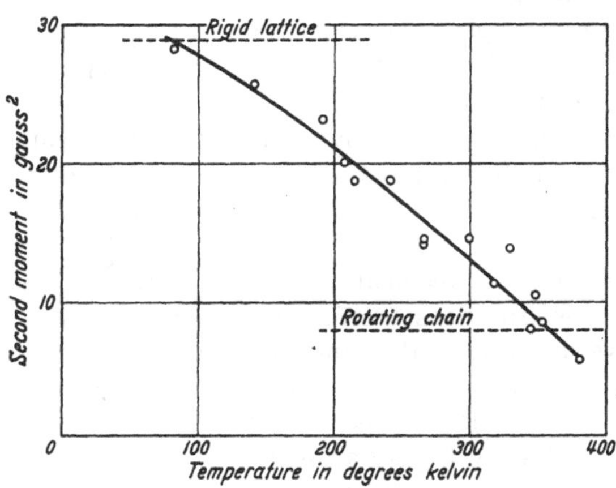

Fig. 8. Variation of second moment with temperature for polyhexamethylene sebacamide [SLICHTER (*78*)]

SLICHTER has calculated ΔH_2^2 for motionless polyamide molecules as a function of the length of the paraffinic segments separating the amide groups. The atomic coordinates used in the calculation are based on X-ray diffraction data [BUNN and GARNER (*16*)] and on conventional bond lengths and angles. Also, following the methods used by ANDREW (*6*) in his calculations on n-paraffins, SLICHTER has calculated ΔH_2^2 for polyamide molecules engaged in classical rotation about the chain axes and also for rotational oscillation. Fig. 8 shows the measured variation of ΔH_2^2 with temperature for polyhexamethylene sebacamide, and also the theoretical values for the rigid lattice and for rotation of all the chains. Although the motion must consist of a composite of different classes, if one considers for simplicity that the motion is solely rotational, then in polyhexamethylene sebacamide all the chain segments are rotating at a temperature of about 360° K, over 100° below the melting point. A crystallographic transition, from the triclinic form to the pseudo-hexagonal, occurs in these polymers over the same region of temperature (SLICHTER (*78*)].

7.2 Polytetrafluoroethylene (PTFE). Nuclear resonance studies of PTFE have been made by SMITH (*82*); NISHIOKA, KOMATSU, and KAKIUCHI (*54*); SLICHTER (*80*); and WILSON and PAKE (*91*). Here it is the fluorine nucleus which is detected, whereas in the other polymers mentioned so far the resonance was that of the proton. Particular interest attaches to the structure and motion of the PTFE molecule, because of some of the unique physical properties of the polymer and because of the room-temperature transitions occurring in many of the properties. Thus, in the neighborhood of 295° K, transitions have been described in the specific volume [RIGBY and BUNN (*72*); QUINN, ROBERTS and WORK (*68*)]; in the crystal structure [BUNN and HOWELLS (*18*)]; in the specific heat [FURUKAWA, MCCOSKEY and KING (*30*); MARX and DOLE (*48*)]; and in the coefficient of friction [FLOM and PORILE (*25*)].

Fig. 9 shows ΔH_2^2 as a function of temperature, measured by SMITH (*82*) on a British PTFE and by SLICHTER (*80*) on two American

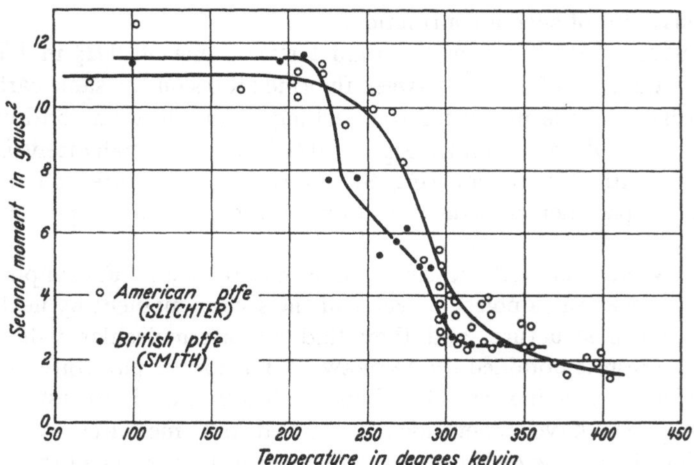

Fig. 9. Variation of second moment with temperature for two American polytetrafluoroethylenes [SLICHTER (*80*)] and for a British polytetrafluoroethylene [SMITH (*82*)]

PTFE samples. The X-ray diffraction studies of BUNN and HOWELLS (*18*) show that the PTFE molecule is helical in structure. Calculation of the intramolecular contribution to ΔH_2^2 is therefore somewhat complicated, yet it can be done with an accuracy which is set only by the uncertainty in the value assumed for the $F—C—F$ angle and the $C—F$ distance. However, even though the crystallographic unit cell is known, one may only speculate as to how the fluorine atoms on one chain are disposed with respect to those on an adjacent chain. SMITH has taken two models of the chain packing and has calculated approximate values of the intermolecular contribution to the total ΔH_2^2. Subtracting these figures from the experimental value at 77° K, he arrives at two values for the intramolecular ΔH_2^2. Comparing these with values calculated for the helical chain, he adduces that the $F—F$ distance for fluorine atoms bound to the same carbon atom lies between 2.08 and 2.18 Å. Using 1.36 Å as the $C—F$ distance, as it has been found to be in some small molecules, SMITH concludes that the $F—C—F$ angle in PTFE must be several degrees smaller than the tetrahedral angle. This sort of structural determination, in systems which are difficult or even impossible to treat by the older methods, shows the special value of the nuclear resonance approach. It should be recalled, though, that the nuclear resonance result is set by internuclear separations. The determination of the bond angle therefore depends upon the accuracy with which the associated bond distance is known, and vice versa. In the present case, the finding of a contraction in the $F-C-F$ angle is plausible, but still there is enough variation in the microwave data for the $C-F$ distance [GORDY, SMITH and TRAMBARULO (*31*)] in small molecules that it is premature to be convinced of such a contraction.

SLICHTER's calculation of the rigid lattice value of ΔH_2^2 in PTFE assumes a distance of 2.11 Å between fluorine atoms on the same carbon. Postulating a simple model for the packing of fluorines on continuous chains, he calculates a total ΔH_2^2 of 10.0 Gauss^2, in moderately good agreement with experiment (Fig. 9). However, there are too many independent parameters in the model for a critical test of the postulated structure.

WILSON and PAKE (*90*) measure a value of 12.0 Gauss^2. For simplicity, they compute an intramolecular value of the second moment by neglecting the helical structure, and they find an intermolecular value by scaling the results obtained by ANDREW (*3*) for normal paraffins. Their total calculated value is 11.7 Gauss^2. NISHIOKA, KOMATSU, and KAKIUCHI (*54*) have independently carried out measurements and calculations like those of WILSON and PAKE, and have reached the same results.

ΔH_2^2 for rotating chains has been calculated by SMITH and by SLICHTER; the value is about 2.5 Gauss^2, in close agreement with the measured value above room temperature. SMITH's experiments show two temperature regions in which ΔH_2^2 decreases with increasing temperature. One transition, occurring between 210 and 230° K, yields a decrease in ΔH_2^2 from the rigid-lattice value to an intermediate value of 6.0 Gauss^2. SMITH calculates that this figure corresponds to the rotation of two-thirds of the chains. He recalls that ANDREW (*6*) found similar behavior in certain long-chain n-paraffins and explained it with the postulate that the chains are meshed like a set of gears, so that rotation must involve just two-thirds of them. The notion that markedly different classes of motion should belong, in an orderly way, to adjacent chains which are structurally equivalent is needlessly artificial. Even if the model were to have some slight merit with n-paraffins, though, one would not expect it to apply with PTFE; for studies already cited in X-ray diffraction and in measurement of frictional coefficient indicate that the PTFE molecule is cylindrical in conformation, with very little tendency to be meshed with its neighbors in the unit cell.

The transition seen by SMITH in the region of 210—230° K does not correspond to any known transition in structure or physical properties. The PTFE's studied by SLICHTER show only one transition in ΔH_2^2, in the neighborhood of room temperature, a transition which also is seen by SMITH (Fig. 9). The onset of motion seen here corresponds in temperature to transitions found in specific volume, crystal structure, specific heat, and frictional coefficient (see above). BUNN and HOWELLS (*18*), examining the crystallographic change, suggest that it may be due either to rotation of the chains or else to translations in the direction of the chain axes. SLICHTER has measured ΔH_2^2 for PTFE fiber at 300° K, with the fiber axis at chosen angles with respect to the magnetic field. The variation of ΔH_2^2 with orientation angle in the field is what would be expected if the PTFE molecules are rotating about the chain axes.

SLICHTER has suggested that the low temperature transition found by SMITH corresponds to the motion of an impurity, perhaps a low molecular weight fraction of the PTFE, which is absent in the polymers studied by SLICHTER. There may also have been important differences between the samples with respect to degree of crystallinity, or in molecular structure. Chain branching is not known to be a feature of PTFE as it is in polyethylene; but still there are evidently differences on a molecular scale among PTFE's, for the two samples studied by SLICHTER showed somewhat different line widths over much of the temperature range (though the differences are within the experimental scatter in ΔH_2^2 data). Both of these polymers were made by the same manufacturer, but they bear different trade names. Whatever may be the sources of the differences among PTFE's, it is important to emphasize that chemical purity often is highly significant in NMR studies. The hazard in NMR work of being misled by mobile impurity molecules has been discussed by RUSHWORTH (*75*).

7.3 Fluorine Derivatives of Polyethylene. The obvious differences in the physical properties of polyethylene and PTFE have suggested the study of the proton and fluorine resonances of the partially-fluorinated derivatives of polyethylene: polyvinyl fluoride, polyvinylidene fluoride, and polytrifluoroethylene [SLICHTER (*80*)]. In these compounds, all of which melt at about 475° K, the onset of chain motion as shown by narrowing of the resonance occurs over a broad temperature range, between about 200 and 350° K. The temperature at which the narrowing is ultimately achieved is higher than in some nonpolar polymers, but it is generally lower than in the polyamides. Evidently the polar groups in the partially-fluorinated derivatives of polyethylene offer only moderate restraint to the rotational motions of the chains, for the concentration of polar groups in these polymers is much greater than in the polyamides. The glass transition has been reported for only one of these polymers, polyvinylidene fluoride, for which T_g is about 238° K [MANDELKERN, MARTIN and QUINN (47)].

7.4 Polyethylene (PE). It is well known that polyethylenes differ widely in many physical properties, owing chiefly to dissimilarities in

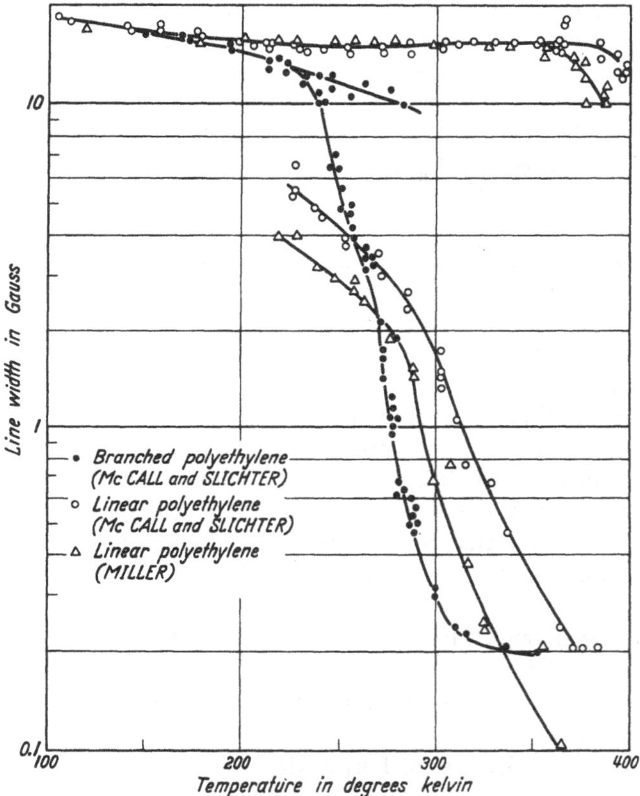

Fig. 10. NMR line with as a function of temperature for a branched polyethylene [MCCALL and SLICHTER (49)] and for a linear polyethylene (separate measurements by MCCALL and SLICHTER and by MILLER (50))

molecular weight and chain branching [RICHARDS (70)]. In particular, there are marked contrasts between PE's made at high pressure, which exhibit extensive chain branching, and those made at low pressure by special catalysis, which contain little if any chain branching.

Fig. 10 shows the line width as a function of temperature for a branched PE and a linear PE, from the studies of MCCALL and SLICHTER (49). In each case, over much of the temperature range, the absorption evidently consists of two superposed curves, one narrow and one broad. These components are ascribed [WILSON and PAKE (90)] to the amorphous and the crystalline portions of the polymer, respectively. The broad component for the linear polymer changes very little with temperature until shortly below the melting point, which occurs at about 410° K. The broad component for the branched polymer becomes progressively narrower with increasing temperature, and ceases to be detectable shortly above room temperature. This disappearance occurs partly because of the much greater intensity of the narrow curve, but also presumably because of the onset of melting of the crystallites, even near room temperature; and because the increased motion of segments in the crystallites at higher temperatures causes the broader curve to blend into the wings of the narrower curve. These observed differences between the crystalline regions of the two types of polymers are in keeping with specific volume studies [MANDELKERN, HELLMAN, BROWN, ROBERTS, and QUINN (*46*)], which show that most of the melting of polymethylene, essentially a linear PE, occurs within a few degrees of the ultimate melting point, whereas the melting of a branched PE occurs over a broad temperature range. MCCALL and SLICHTER conclude that there is a marked difference in the chain motion within crystallites of the branched polymer compared to the linear, and they ascribe the difference to the existence of lattice defects in the arrays of the branched material, perhaps from the incorporation of chain branch points into the crystallites.

Comparing the narrow components of the absorption for the two types of PE, MCCALL and SLICHTER note that the line narrowing occurs at a much lower temperature in the branched polymer than in the linear (Fig. 10). This behavior is taken to support the evidence from X-ray diffraction [SLICHTER (*79*)] that the high degree of crystallinity in linear PE's imposes constraints upon the "amorphous" regions. The very small line widths achieved by the narrow curves of these substances, below the gross melting point in each case and even below room temperature for the branched PE, are taken to signify the existence of vigorous molecular motion, comparable to that found with some viscous liquids. At low temperatures, on the other hand, the NMR evidence of the freezing of motion in the amorphous portions supports the view [CAREY, SCHULZ and DIENES (*19*)] that the shear modulus in PE's is dominated at low temperatures, of the order of 170° K and below, by the rigidity of the chain segments in the amorphous portions.

Branched and linear PE's have also been studied by MILLER (*50*). He reports broad resonances comparable to those found by MCCALL and SLICHTER, but he states that the narrow curves approach asymptotic values at low temperature which are much less than the line widths for the crystalline values. MILLER's data for the same type of linear polymer as that studied by MCCALL and SLICHTER are also shown in Fig. 10. In measuring the width of the narrow component, MILLER has sought to take account of the fact that the observed line width of the narrow component is affected to some degree by the broad component. McCALL and SLICHTER recognize this difficulty, but they feel that the nature and amount of this influence are ambiguous, and so they have graphed the observed line widths, without any correction. Hence the differences shown in Fig. 10 between the two curves for the linear PE stem from the interpretation, not from the polymer.

Contrary to the data of McCALL and SLICHTER, MILLER reports no transition region in which the narrower component simply broadens into the absorption curve for the crystalline regions. MILLER postulates that in the temperature range where the narrow component becomes evident, near 225° K, there is a fairly sudden onset of motion among the chain segments in the amorphous regions, perhaps corresponding to a second-order transition. However, it is difficult to distinguish the narrower component accurately at the low temperatures where it first appears. Hence it is risky to judge that the narrower component tends toward an asymptotic value separate from the broad component, instead of becoming progressively wider at very low temperatures. Moreover, it seems improbable that at these low temperatures there could be a rather abrupt onset of chain motion, sufficient to reduce the line width to a third of its rigid lattice value. It is more reasonable to suppose that there is a gradual increase in the motion in the amorphous regions over a rather broad temperature range.

As with other substances, comparisons between experiment and postulated models of the structure and motion rely upon ΔH_2^2. McCALL and SLICHTER find that the calculated value of ΔH_2^2 for the rigid lattice, taking into account the contraction of the lattice at low temperatures, lead to figures which are significantly less than the measured values. The calculations assume that the $C-H$ bond distance has the familiar value, 1.09 Å, obtained from spectroscopic studies of gaseous molecules, and that the $H-C-H$ angle is tetrahedral. McCALL and SLICHTER conclude that the disparity between model and experiment cannot be reconciled unless the proton-proton distance in the CH_2 group is reduced from 1.79 Å, as given by these bond parameters, to about 1.70 Å. It is of course impossible to say, from NMR data alone, whether the contraction must be sought in the bond angle or the distance. It should be remarked that R. L. MILLER and also R. L. COLLINS (private communications to the author) are skeptical about the reported disparity between experiment and calculation, and suggest that the experimental results may have been affected at low temperatures by the phenomenon of saturation, which we have already mentioned. McCALL and SLICHTER feel that they largely avoided this difficulty, but no doubt the matter deserves more study.

Calculation of ΔH_2^2 for rotating chains of PE leads to values which are well above those found even below room temperature. Indeed, McCall and Slichter show that even if the individual methylene groups in PE are presumed to rotate independently of one another, a condition which cannot obtain in the solid without disintegration of the chains, the resulting figure calculated for ΔH_2^2 lies above the values observed near room temperature. Hence the amorphous regions of PE must contain, well below the ultimate melting point, considerable translational motion, as well as rotational motion.

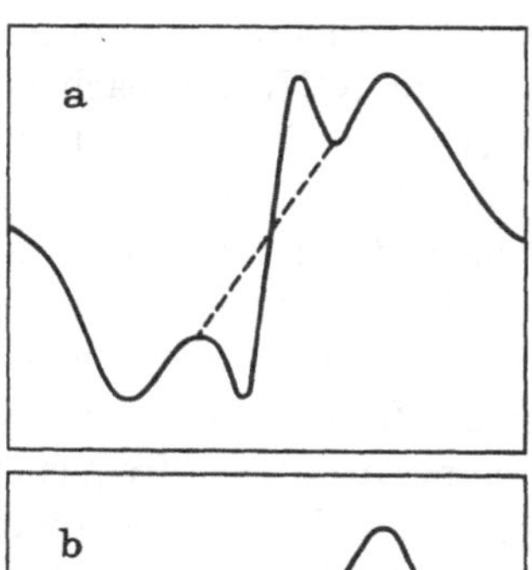

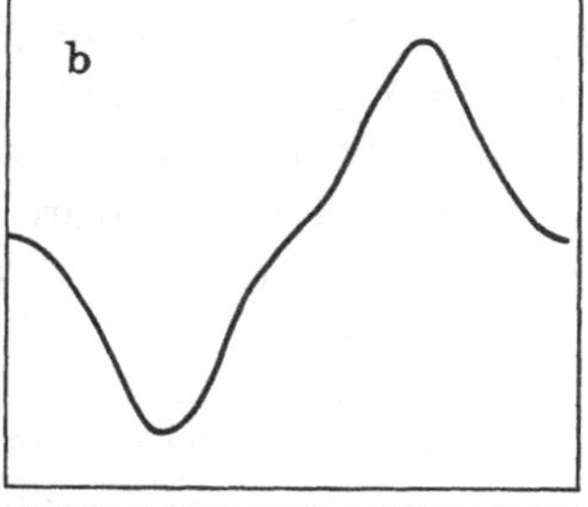

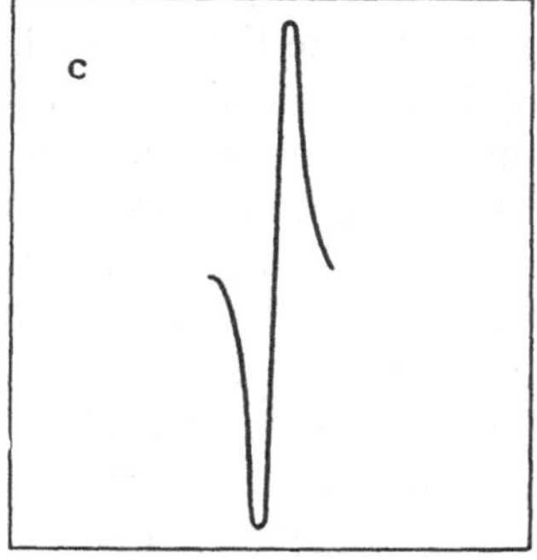

Fig. 11. Variation of shape of derivative curve (schematic): (a) compound shape, with broad and narrow parts separated graphically; (b) typical shape at low temperature; (c) typical shape at high temperature [Slichter and McCall (*87*)]

7.5 Degree of Crystallinity. One of the most important parameters used in characterizing polymers is the degree of crystallinity. A number of studies of this quantity have been made by various methods, principally using measurements of density or of comparative intensity of X-ray diffraction from crystalline and amorphous phases. Wilson and Pake (*90*) have noted that in samples of PE and PTFE, the NMR curve consists of a broad and a narrow component, ascribed to the crystalline and the amorphous regions of the polymers, respectively. Measurement of the ratio of the area under the broad curve to the area under the total absorption curve should describe the degree of crystallinity. Actually, since one ordinarily observes experimentally the derivative of the absorption curve, the area under the absorption is proportional to the first moment of the derivative curve. The procedure involves a graphical separation of the derivative curve into a broad constituent and a narrow constituent (Fig. 11 a). This graphical decomposition is usually done with a straight line, or with some reasonable curve. But any method of decomposition is arbitrary and inherently causes some uncertainty. Thus, although very good reproducibility can be attained, and quite close agreement with other types of crystallinity measurement can be achieved [Collins (*21*)], one is really not entitled to express NMR crystallinity determinations to an accuracy of a few per cent. Indeed, since semicrystalline polymers are not two-phase systems anyhow, but rather are systems possessing a broad and continuous range of the degree of order, fine comparisons of the extent of crystallinity determined by whatever method are unrealistic.

The NMR approach is not universally applicable. SLICHTER and McCALL (*81*) point out that a basic criterion is that motions in the amorphous portion occur at frequencies large compared with the frequency for line narrowing, about 10^4 c. p. s., while at the same temperature motions in the crystalline regions occur at frequencies which are small on this scale. At low temperatures, the frequencies in both regions are insignificant, and the derivative curve appears as in Fig. 11b; at high temperatures, the frequencies in both regions are much larger than 10^4 c. p. s., and the curve appears as in Fig. 11c. Although a curve such as that in Fig. 11a permits the crystallinity measurement, it does not follow that the necessary set of conditions will be met with every polymer, regardless of temperature. For example, the compound structure of Fig. 11a does not appear at any temperature in Hevea rubber [GUTOWSKY and MEYER (*35*)] or in the partially-fluorinated derivatives of polyethylene [SLICHTER (*80*)]. Even when the compound curve is found, agreement with the results of other kinds of crystallinity measurement may hold only over a limited range of temperature. SLICHTER and McCALL note that with a linear PE the degree of crystallinity, as found by NMR, changes little with temperature over a broad span of temperature, and agrees well with data from other methods. With a branched PE, though, the NMR result changes rapidly with temperature, and agrees with data from other methods over only a short interval of temperature. Indeed, the vanishing of the broad component above room temperature (see above) precludes study by NMR of the degree of crystallinity in branched PE's at higher temperatures. Evidently caution must be exercised in the use of NMR for crystallinity determinations. It is necessary to establish that there is a correspondence between degree of crystallinity, which depends on the morphology, and the NMR measurement, which depends on the vigor of motion. An example of the difficulty may be found in the study of polyethylene which has been irradiated at high energy; here the irradiation both diminishes the crystallinity and introduces constraints to chain motion, in a fashion which depends upon irradiation dosage. It seems likely that attempts to compare the degree of crystallinity at widely different dosages, using the NMR measurement of crystallinity [FUJIWARA, AMAMIYA and SHINOHARA (*28*)], will prove to be unreliable. Therefore, although NMR is helpful in selected cases for determination of the extent of crystallinity, the results must be examined closely in terms of the findings from other methods, for *each* polymer under consideration.

8. Thermal Activation of Molecular Motion

The thermal activation of molecular motion in polymers and other substances has been investigated in studies of mechanical and dielectric relaxation, and recently by NMR. It is well known, from mechanical and dielectric data, that polymeric systems are characterized by broad distributions in relaxation times. Although the spectrum of relaxation times in a polymer at a given temperature is ordinarily quite complicated, the relaxation behavior of a number of polymers has been examined in great detail over wide ranges of times and temperatures. Less detailed investigations of the temperature dependence of mechanical and electrical relaxation processes involve the simplifying assumption that the motional process can be described in terms of a single, characteristic relaxation time. Application of this assumption to a wide variety of systems has yielded reasonable results and has led to acceptance of the procedure. One must recall, though, that the activation energy thus obtained represents a complicated average over the spectrum of motions existing in the system under study. Furthermore, since the form of the distribution curve for relaxation varies with the substance and with the method of measurement (mechanical versus electrical), the averaging process is not strictly equivalent from one case to another. Hence detailed comparisons of activation energies are unjustified, even though they are often tempting.

Activation energies and correlation (relaxation) frequencies may be found from NMR data through the relations [GUTOWSKY and PAKE (*33*)]

$$2\pi \nu_c = \alpha \gamma \, \delta H / \tan \left[(\pi/2) \, (\delta H^2 - B^2)/C^2\right] \tag{8}$$

and

$$\nu_c = \nu_0 \exp(-E/RT) \, . \tag{9}$$

Here ν_c is the correlation frequency for the motion taken to be responsible for the narrowing of the resonance envelope, and ν_0 and E are the frequency factor and activation energy for the motion. δH is the line width in the transition region, B is the line width at a temperature above the transition region, and C is the line width below this region. The constant α may be taken [KUBO and TOMITA (*43*)] to be 0.18. The constant γ is the gyromagnetic ratio, and is equal to $2\pi\mu/I\,h$, where h is the Planck constant and where μ and I have the meanings given in Section 2.3.

The nature of the averaging process which results from the use of a single correlation frequency, ν_c, is obscure, but the approach is doubtless analogous to what is involved in treatments of other forms of relaxation data. An additional uncertainty comes from the use of NMR line widths to evaluate the activated process, for one is compelled to overlook the fact that the intermolecular contributions to the resonance curve vary with the motion, and hence with the temperature, in a different fashion from the intramolecular. Second moments may be used instead of line widths, in an expression comparable to Equation (8) [POWLES and GUTOWSKY (65)], but again some error enters from the non-parallel

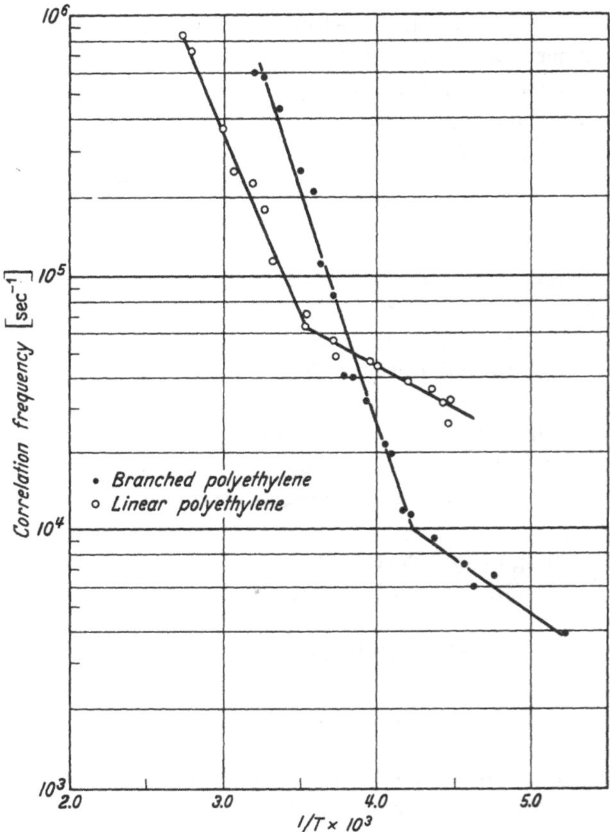

Fig. 12. Variation of correlation frequency with inverse temperature for a linear polyethylene and a branched polyethylene [MCCALL and SLICHTER (49)]

variation of the intermolecular and intramolecular contributions over the temperature range. Despite these difficulties, fruitful comparisons may be made among NMR results, and with activation energies found by other means.

Fig. 12 gives an example of a graph of $\ln \nu_c$ plotted against $1/T$, in conventional fashion, and used to find E and ν_0. The data, from the work of MCCALL and SLICHTER (49), come from the line width studies on a linear PE and a branched PE shown above in Fig. 12. The values of E and ν_0 computed from a number of NMR studies on polymers are given in Table 1. It should be emphasized that these values are highly approximate. For example, in assessing the line-width data for the linear PE, McCall and Slichter can estimate reasonable values for B and C [Equation (8)] which lead to an activation energy of as much as 12 kcal./mole for the high-temperature process, a figure twice as great as that given in Table 1.

Table 1. *Activation Energy and Frequency Factor for Motional Narrowing of Resonance Line in Several Polymers*

Polymer	E [kcal/mole]	ν_0 [sec^{-1}]	Reference
Polyethylene, branched (low-temperature narrowing)	2.0	6.6×10^5	49
Polyethylene, branched (high-temperature narrowing)	8.0	2.6×10^{11}	49
Polyethylene, linear (low-temperature narrowing)	1.5	8.7×10^5	49
Polyethylene, linear (high-temperature narrowing)	6.1	5.0×10^9	49
Polyethylene, linear	8.2	—	22
Natural rubber, cured (low-temperature narrowing)	4.5	3.2×10^9	35
Natural rubber, cured (high-temperature narrowing)	13.3	6.0×10^{16}	35
Polyisobutylene	10.2	1.9×10^{13}	66
Polyisobutylene, swelled with benzene	12.6	4.2×10^{16}	66
Polyvinyl fluoride (proton resonance)	2.3	2.2×10^6	80
Polyvinylidene fluoride (proton resonance)	1.8	1.0×10^6	80
Polytrifluoroethylene (proton resonance)	2.6	6.3×10^6	80
Polytetrafluoroethylene	4.3	8.9×10^7	80

It is natural to seek comparisons between these NMR data and the findings of other methods. Studies on PE are numerous. McCall and Slichter have compared the NMR results, belonging to the amorphous regions and characterized by activation energies of 6—8 kcal./mole or somewhat more, with results from studies on mechanical and dielectric relaxation, and in permeation by small molecules. The activation energies from these other studies, in which chain motion likewise is ascribed to the amorphous regions, lie in the range of 10—16 kcal./mole. The NMR results are in reasonable agreement with these other methods.

9. Conclusion

Although nuclear magnetic resonance spectroscopy is a comparatively new field, it has already contributed in important ways to our knowledge of molecular structure and motion in many substances, including some important groups of high polymers. In dealing with high polymers, NMR suffers from some of the same difficulties encountered by other methods of measurement, such as the difficulties ensuing from the broad distribution of molecular weights and from the mesomorphic structures of many polymers. We have remarked on some incorrect conclusions coming from some of the early NMR work on polymers. Although some work needs to be repeated, a more important need is for systematic, quantitative comparisons between closely related members of a polymeric series, such as a group of related copolymers, rather than between isolated examples of polymers. For instance, our understanding of chain stiffness and structure would be aided by a study of a series of copolymers from vinyl chloride and vinylidene chloride; similarly, our understanding of elastic behavior would be helped by a study of a series of styrene-butadiene copolymers. With comparable studies in other systems, important features such as reinforcement, vulcanization, and interaction of electric dipoles would become very much better defined. There are other aspects of NMR studies which have been largely or totally overlooked. For example, investigations so far have been limited to the resonances of the hydrogen and fluorine nuclei, but there are instances in which the study of other nuclei would be valuable. Examples of such nuclei are Si^{29}, in suitably enriched samples; and nuclei possessing more than two energy states, such as N^{14} and the isotopes of chlorine. The selective labeling of different parts of the molecule with protons and with deuterium (or tritium) would allow distinction between separate modes of motion in certain cases. Effects of cross-linking induced by ionizing radiation deserve attention. The action of plasticizers upon the motion of polymer molecules, and the state of association between plasticizer and polymer, have been studied only in preliminary fashion. Variables such as pressure should be exploited, and NMR techniques such as the pulse methods (see above) should be called upon.

There are, then, a great many areas in which NMR may be expected to make important contributions to our understanding of polymers. It would be well to point out once more, though, that the NMR approach is far from being self-sufficient; rather, the emphasis must be upon combining the NMR results with those from other methods.

The writer wishes to express his gratitude to Dr. D. W. McCALL for many stimulating discussions.

References

1. ALFORD, S., and M. DOLE: Specific heat of synthetic high polymers. VI. A study of the glass transition in polyvinyl chloride. J. Amer. chem. Soc. **77**, 4774—4777 (1955).
2. ALPERT, N. L.: Studies of the solid state by means of nuclear magnetism. Phys. Rev. **72**, 637—638 (1947).
3. — Study of phase transitions by means of nuclear magnetic resonance phenomena. Phys. Rev. **75**, 398—410 (1949).

4. Anderson, H. L.: Precise measurement of the gyromagnetic ratio of He^3. Phys. Rev. **76**, 1460—1470 (1949).
5. Andrew, E. R., and R. Bersohn: Nuclear magnetic resonance line shape for a triangular configuration of nuclei. J. chem. Physics **18**, 159—161 (1950).
6. — Molecular motion in certain solid hydrocarbons. J. chem. Phys. **18**, 607—618 (1950).
7. —, and R. G. Eades: A nuclear magnetic resonance investigation of solid cyclohexane. Proc. roy. Soc. (Lond.) A **216**, 398—412 (1953).
8. — — A nuclear magnetic resonance investigation of three solid benzenes. Proc. roy. Soc. (Lond.) A **218**, 537—552 (1953).
9. — Nuclear magnetic resonance. Cambridge: The University Press 1955.
10. Baker, W. O., and C. S. Fuller: Macromolecular disorder in linear polyamides. Relation of structure to physical properties of copolyamides. J. Amer. chem. Soc. **64**, 2399—2407 (1942).
11. Banas, E. M., B. A. Mrowca and E. Guth: Nuclear magnetic relaxation in natural and synthetic rubbers. Phys. Rev. **98**, 265 (1955).
12. — — Nuclear spin-spin relaxation time in polymers. Phys. Rev. **98**, 1548 (1955).
13. Bloch, F., W. W. Hansen and M. Packard: The nuclear induction experiment. Phys. Rev. **70**, 474—485 (1946).
14. Bloembergen, N., E. M. Purcell and R. V. Pound: Relaxation effects in nuclear magnetic resonance absorption. Phys. Rev. **73**, 679—712 (1948).
15. Bunn, C. W.: Molecular structure and rubber-like elasticity. III. Molecular movements in rubber-like polymers. Proc. roy. Soc. (Lond.) A **180**, 82—99 (1942).
16. —, and E. V. Garner: The crystal structures of two polyamides ("Nylons"). Proc. roy. Soc. (Lond.) A **189**, 39—68 (1947).
17. — The study of high-polymer structure by X-ray diffraction methods. J. chem. Soc. **1947**, 297—306.
18. —, and E. R. Howells: Structure of molecules and crystals of fluorocarbons. Nature (Lond.) **174**, 549—551 (1954).
19. Carey, R. H., E. F. Schulz and G. J. Dienes: Mechanical properties of polyethylene. Ind. Eng. Chem. **42**, 842—847 (1950).
20. Carr, H. Y., and E. M. Purcell: Effects of diffusion on free precession in nuclear magnetic resonance experiments. Phys. Rev. **94**, 630—638 (1954).
21. Collins, R. L.: Crystallinity of polyethylene by nuclear resonance. Bull. Amer. phys. Soc. **1**, 216 (1956).
22. — Nuclear spin resonance line width measurements of polyethylene. Bull. Amer. phys. Soc. **2**, 103 (1957); J. Polymer Sci. **27**, 67—74 (1958).
23. Deutsch, K., E. A. W. Hoff and W. Reddish: Relation between structures of polymers and their dynamic mechanical and electrical properties. I. Some alpha-substituted acrylic ester polymers. J. Polymer Sci. **13**, 565—582 (1954).
24. Evans, A. G., and M. Polanyi: Steric hindrance and heats of formation. Nature (Lond.) **152**, 738—740 (1943).
25. Flom, D. G., and N. T. Porile: Friction of teflon sliding on teflon. J. appl. Phys. **26**, 1088—1092 (1955).
26. Flory, P. J.: Principles of polymer chemistry. p. **458**. Ithaca: Cornell University Press 1953.
27. Fox, T. G., and P. J. Flory: Second-order transition temperatures and related properties of polystyrene. I. Influence of molecular weight. J. appl. Phys. **21**, 581—591 (1950).
28. Fujiwara, S., A. Amamiya and K. Shinohara: Nuclear magnetic resonance in irradiated polyethylene. J. chem. Phys. **26**, 1343 (1957).

29. FULLER, C. S., C. J. FROSCH and N. R. PAPE: X-ray examination of polyisobutylene. J. Amer. chem. Soc. **62**, 1905—1913 (1940).
30. FURUKAWA, G. T., R. E. MCCOSKEY and G. J. KING: Calorimetric properties of polytetrafluoroethylene (teflon) from 0 to 365° K. J. Res. nat. Bur. Standards **49**, 273—278 (1952).
31. GORDY, W., W. V. SMITH and R. F. TRAMBARULO: Microwave spectroscopy. pp. 371—373 (References). New York: D. Van Nostrand Company, Inc. 1953.
32. GUTOWSKY, H. S., G. B. KISTIAKOWSKY, G. E. PAKE and E. M. PURCELL: Structural investigations by means of nuclear magnetism. I. Rigid crystal lattices. J. chem. Phys. **17**, 972—981 (1949).
33. —, and G. E. PAKE: Structural investigations by means of nuclear magnetism. II. Hindered rotation in solids. J. chem. Phys. **18**, 163—170 (1950).
34. —, L. H. MEYER and R. E. MCCLURE: Apparatus for nuclear magnetic resonance. Rev. sci. Instr. **24**, 644—652 (1952).
35. — — The proton magnetic resonance in natural rubber. J. chem. Phys. **21**, 2122—2126 (1953).
36. —, A. SAIKA, M. TAKEDA and D. E. WOESSNER: Proton magnetic resonance studies on natural rubber. II. Line shape and T_1 measurements. J. chem. Phys. **27**, 534—542 (1957).
37. HAHN, E. L.: Spin echoes. Phys. Rev. **80**, 580—594 (1950).
38. HOLROYD, L. V., R. S. CODRINGRON, B. A. MROWCA and E. GUTH: Nuclear magnetic resonance study of transitions in polymers. J. appl. Phys. **22**, 696 to 705 (1951).
39. HONNOLD, V. R., F. MCCAFFREY and B. A. MROWCA: Studies of rubber-like polymers by nuclear magnetism. J. appl. Phys. **25**, 1219—1223 (1954).
40. JENCKEL, E.: Die Einfriertemperatur hochmolekularer Gläser und ihr chemischer Aufbau. Kolloid-Z. **100**, 163—170 (1942).
41. KNAPPE, W., and A. SCHULZ: Die refraktometrische Bestimmung der Wirksamkeit von Weichmachern bei Polyvinylchlorid. Kunststoffe **41**, 321—324 (1951).
42. KOJIMA, S., and S. OGAWA: Proton magnetic resonance absorption in cetyl alcohol. J. phys. Soc. Japan **8**, 283—287 (1953).
43. KUBO, R., and K. TOMITA: A general theory of magnetic resonance absorption. J. phys. Soc. Japan **9**, 888—919 (1954).
44. LIQUORI, A. M.: Molecular configurations of stretched polyisobutylene. Acta crystallogr. (Lond.) **8**, 345—347 (1955).
45. LOWE, I. J., L. O. BROWN and R. E. NORBERG: Nuclear magnetic relaxation times in polyethylene. Bull. Amer. phys. Soc. **30**, 16 (1955).
46. MANDELKERN, L., M. HELLMAN, D. W. BROWN, D. E. ROBERTS and F. A. QUINN JR.: The melting transition of polymethylene. J. Amer. chem. Soc. **75**, 4093—4094 (1953).
47. —, G. M. MARTIN and F. A. QUINN JR.: Glass temperatures of polychlorotrifluoroethylene, polyvinylidene fluoride, and their copolymers. Bull. Amer. phys. Soc. **1**, 123 (1956); J. Res. nat Bur. Standards **58**, 137—143 (1957).
48. MARX, P., and M. DOLE: Specific Heat of synthetic high polymers. V. A study of the order-disorder transition in Polytetrafluoroethylene. J. Amer. chem. Soc. **77**, 4771—4774 (1955).
49. MCCALL, D. W., and W. P. SLICHTER: Molecular motion in polyethylene. J. Polymer Sci. **26**, 171—186 (1957); Bull. Amer. phys. Soc. **2**, 125 (1957).
50. MILLER, R. L.: The nuclear magnetic resonance of polymers. Bull. Amer. phys. Soc. **2**, 125 (1957); R. C. REMPEL, H. E. WEAVER, R. H. SANDS and R. L. MILLER; Nuclear magnetic resonance studies of polyethylene. J. appl. Phys. **28**, 1082—1089 (1957).

51. Mrowca, B. A., L. V. Holroyd and E. Guth: Study of high polymers by nuclear magnetism. II. Line widths through transition temperatures. Phys. Rev. **79**, 1026—1027 (1950).
52. Newman, R.: Proton magnetic resonance in polyethylene. J. chem. Phys. **18**, 1303—1304 (1950).
53. Nielsen, L. E., R. Buchdahl and G. C. Claver: Molecular structure of styrene-butadiene copolymers. Dynamic mechanical measurements. Ind. Eng. Chem. **42**, 341—345 (1951).
54. Nishioka, A., H. Komatsu and Y. Kakiuchi: Nuclear magnetic resonance in some crystalline polymers. J. phys. Soc. Japan **12**, 283—285 (1957).
55. Nolle, A. W.: Nuclear magnetic resonance relaxation times for polyisobutylene in carbon tetrachloride solution. Phys. Rev. **98**, 1560 (1955).
56. Odajima, A., J. Sohma and M. Koike: Line-width transition of the proton magnetic resonance in polymers. J. chem. Phys. **23**, 1959—1960 (1955).
57. — — — Proton magnetic resonance in chain polymers. J. phys. Soc. Japan **12**, 272—282 (1957).
58. Oshima, R., and H. Kusumoto: Effect of elongation on the proton magnetic resonance of natural rubber. J. chem. Phys. **24**, 913 (1956).
59. Pake, G. E.: Nuclear resonance absorption in hydrated crystals: Fine structure of the proton line. J. chem. Phys. **16**, 327—336 (1948).
60. — Fundamentals of nuclear magnetic resonance absorption. Amer. J. Phys. **18**, 438—452, 473—486 (1950).
61. — Remarks reported in Disc. Faraday Soc. **19**, 252 (1955).
62. Patnode, W., and W. J. Scheiber: The density, thermal expansion, vapor pressure, and refractive index of styrene, and the density and thermal expansion of polystyrene. J. Amer. chem. Soc. **64**, 3449—3451 (1939).
63. Pound, R. V., and W. D. Knight: A radiofrequency spectrograph and simple magnetic field meter. Rev. sci. Instr. **21**, 219—225 (1950).
64. Powles, J. G., and H. S. Gutowsky: Proton magnetic resonance of the CH_3 group. I. Investigation of six tetrasubstituted methanes. J. chem. Phys. **21**, 1695 to 1703 (1953).
65. — — Proton magnetic resonance of the CH_3 group. III. Reorientation mechanism in solids. J. chem. Phys. **23**, 1692—1699 (1955).
66. — Nuclear magnetic resonance absorption in polyisobutylene. Proc. phys. Soc. (Lond.) **69**, 281—292 (1956).
67. — Nuclear magnetic resonance absorption in polymethyl methacrylate and polymethyl α-chloroacrylate. J. Polymer Sci. **22**, 79—93 (1956).
68. Quinn jr., F. A., D. E. Roberts and R. N. Work: Volume-temperature relationship for the room temperature transition in teflon. J. appl. Phys. **22**, 1085—1086 (1951).
69. Rehner jr., J.: Heat conduction and molecular structure in rubber-like polymers. J. Polymer Sci. **2**, 263—274 (1947).
70. Richards, R. B.: Polyethylene-structure, crystallinity, and properties. J. appl. Chem. **1**, 370—376 (1951).
71. Richards, R. E., and J. A. S. Smith: Nuclear magnetic resonance spectra of some acid hydrates. Trans. Faraday Soc. **47**, 1261—1274 (1951).
72. Rigby, H. A., and C. W. Bunn: A room-temperature transition in polytetrafluoroethylene. Nature (Lond.) **164**, 583 (1949).
73. Rochow, E. G., and H. G. Leclair: On the molecular structure of methyl silicone. J. inorg. nucl. Chem. **1**, 92—111 (1955).
74. Rogers, S. S., and L. Mandelkern: Glass formation in polymers. I. The glass transitions of poly-(n-alkyl methacrylates). J. phys. Chem. **61**, 985—990 (1957).

75. RUSHWORTH, F. A.: Nuclear magnetic resonance absorption in anthracene. J. chem. Phys. **20**, 920—921 (1952).
76. SCHILDKNECHT, C. E.: Vinyl and related polymers. pp. 142—145. New York: John Wiley and Sons, Inc. 1952.
77. SCHMIEDER, K., and K. WOLF: Mechanische Relaxationserscheinungen an Hochpolymeren (Beziehungen zur Struktur). Kolloid-Z. **134**, 149—189 (1953).
78. SLICHTER, W. P.: Proton magnetic resonance in polyamides. J. appl. Phys. **26**, 1099—1103 (1955).
79. — On the morphology of highly crystalline polyethylene. J. Polymer Sci. **21**, 141—143 (1956).
80. — Nuclear magnetic resonance in some fluorine derivatives of polyethylene. J. Polymer Sci. **24**, 173—188 (1957).
81. — and D. W. MCCALL: Note on the degree of crystallinity in polymers as found by nuclear magnetic resonance. J. Polymer Sci. **25**, 230—234 (1957).
82. SMITH, J. A. S.: A nuclear resonance investigation of polytetrafluoroethylene. Disc. Faraday Soc. **19**, 207—215 (1955).
83. TAKEDA, M., and H. S. GUTOWSKY: Proton magnetic resonance of solid sym-tetrachloro and tetrabromoethane. J. chem. Phys. **26**, 577—579 (1957).
84. TANAKA, K., K. YAMAGATA and S. KITTATA: Nuclear magnetic resonance of high polymers (phase transition of P. V. A. FIBERS). Bull. chem. Soc. Japan **29**, 843—844 (1956).
85. UEBERREITER, K.: Über das Einfrieren normaler Flüssigkeiten und Flüssigkeiten mit „fixierter" Struktur wie Kautschuk und Kunstharze. Z. physik. Chem. B **45**, 361—373 (1940).
86. VLECK, J. H. VAN: The dipolar broadening of magnetic resonance lines in crystals. Phys. Rev. **74**, 1168—1183 (1948).
87. WATKINS, G. D., and R. V. POUND: An improved R. F. Spectrometer: *g*-factor ratios Li^7/Li^6 and Cl^{35}/Cl^{37}. Phys. Rev. **82**, 343 (1951).
88. WILEY, R. H., and G. M. BRAUER: Refractometric determination of second-order temperatures in polymers. II. Some acrylic, vinyl halide and styrene polymers. J. Polymer Sci. **3**, 455—461 (1948).
89. — — Specific refractivity-temperature data for polyvinyl acetate and polybutyl acrylate. J. Polymer Sci. **4**, 351—357 (1949).
90. WILSON, C. W., III, and G. E. PAKE: Nuclear magnetic resonance determination of degree of crystallinity in two polymers. J. Polymer Sci. **10**, 503 to 505 (1953).
91. — — Nuclear magnetic relaxation in polytetrafluoroethylene and polyethylene. J. chem. Phys. **27**, 115—122 (1957).
92. WÜRSTLIN, F.: Einfriererscheinungen und chemische Konstitution, chapter in Die Physik der Hochpolymeren, edited by H. A. STUART. pp. 639—672 (References). Berlin: Springer 1955.

Fortschr. Hochpolym.-Forsch., Bd. 1. S. 75—113 (1958)

Fluorine-Containing Polymers. I. Fluorinated Vinyl Polymers with Functional Groups, Condensation Polymers, and Styrene Polymers

By

WILLIAM POSTELNEK, LESTER E. COLEMAN, and ALAN M. LOVELACE

Materials Laboratory, Wright Air Development Center, USAF
Air Research and Development Command, Wright-Patterson AFB, Ohio

Contents

List of Tables

I. Introduction

Polytetrafluoroethylene (Teflon) and polycholortrifluoroethylene (Kel-F) served for many applications during World War II, when urgent requirements for thermally stable, solvent resistant polymeric materials were uppermost. During the years following the war, a great deal of research and engineering effort was expended in the synthesis of various fluorine-containing monomers and subsequent polymerization into materials which would function in a variety of applications under conditions of high temperature and in the present of organic solvents, fuels and oils. Teflon and Kel-F are now materials of considerable commercial importance. More recent developments such as a new fluorocarbon elastomer, "Viton A" (E. I. du Pont de Nemours) and a new fluorine-containing silicone elastomer, "Silastic LS-53" (Dow Corning Corp.) also indicate great commercial utility.

It is the purpose of this paper to review the progress in the field of fluorine-containing polymers, excepting those derived from fluorocarbon olefins and dienes. Previously, very little information has been compiled on the subject of fluorine-containing vinyl polymers with functional groups, condensation polymers and homo-and copolymers of styrenes.

Monomer synthesis and polymerization data are presented for several classes of materials. Tables of monomers are listed with some physical properties, and physical, chemical and mechanical properties of polymers are indicated, wherever possible.

II. Vinyl Polymers with Functional Groups

A. Acrylic Esters

A considerable amount of work has been done on the fluorine-containing acrylates. For purposes of a systematic review it may be conveniently divided into two general areas; acrylates in which the fluorine is contained on the acid moiety of the ester, and the acrylates derived from fluoroalcohols. The latter group has received a more detailed and systematic study and will be discussed later in this section.

1. Synthesis and Polymerization of Fluorinated Acrylic Acids and Their Esters. Only two examples of acrylates containing fluorine on both the acid and alcohol moieties are reported. These esters are included in the table of esters (Table 2).

The synthesis of α-fluoroacrylic acid is reported by HENNE and FOX (*50*) using the following sequence of reactions:

$$CH_2{=}CHCO_2CH_3 \rightarrow CH_2BrCHBrCO_2CH_3 \rightarrow [CH_2{=}CBrCO_2CH_3] \rightarrow CH_2BrCBr_2CO_2CH_3 \rightarrow CH_2BrCBrFCO_2CH_3 \rightarrow CH_2BrCBrFCO_2H \rightarrow CH_2{=}CFCO_2H$$

These workers observed that sublimation of the α-fluoroacrylic acid left a viscous pot residue but no polymerization studies were reported. A similar procedure is discussed in a patent by McGINTY (*74, 75*) for the preparation of methyl-α-fluoroacrylates:

$$CH_2{=}CClCO_2CH_3 \rightarrow CH_2BrCBrClCO_2CH_3 \rightarrow CH_2BrCFClCO_2CH_3 \rightarrow CH_2{=}CFCO_2CH_3$$

This latter procedure will give directly, a better over-all yield of the ester.

McGINTY (*74, 75*) and CRAWFORD (*29*) report that methyl-α-fluoroacrylate will homopolymerize using either heat, light or other catalysts to give a clear transparent solid. The properties of the polymer polymerized with ultraviolet light at 25° are reported to be: specific gravity at 15° of 1.38 and N_D^{20}, 1.4565.

Polymethyl-α-fluoroacrylate was also prepared by ANSPON and BARON (*2*) by polymerization with both light and heat. The polymer obtained was a clear tough transparent solid with a heat distortion temperature of 110° and good thermal stability after two hours exposure at 204°.

CRAWFORD (*29*) effected the transesterification of methyl-α-fluoroacrylate and 2-fluoroethanol to obtain 2-fluoroethyl-α-fluoroacrylate. This monomer is also reported to homopolymerize to give a transparent, low refractive index solid.

The preparation of ethyl-β,β-difluoroacrylate is described in a patent by DICKEY et al. (*37*). The oxidative chlorination of $CF_2ClCH{=}CCl_2$ followed by treatment with ethanol gives $CF_2ClCHClCO_2C_2H_5$ which upon treatment with zinc in ethanol gives $CF_2{=}CHCO_2C_2H_5$. It is also claimed that either heat, light or peroxides will induce copolymerization of the monomer with methylacrylate and methylmethacrylate.

The last remaining ester of this group which has received much attention as a monomer is α-trifluoromethylacrylate. This compound is generally prepared from β,β,β-trifluoro-α-hydroxyisobutyronitrile. The preparation of the trifluoroacetone cyanohydrin has been described by DICKEY (*33, 35*) and DARRALL et al. (*31*). The conversion of the cyanohydrin to either the α-trifluoromethylacrylic acid or its esters has been accomplished by a number of synthetic routes.

DICKEY (*32, 33, 35*) reports that the treatment of trifluoroacetone cyanohydrin with the following reagents will yield the acrylic acid or its esters: a) concentrated sulfuric acid, b) concentrated sulfuric acid and sulfur, c) alkyl hydrogen sulfates and d) thionyl chloride. It should be pointed out that BUXTON, STACEY and TATLOW (*15*) were unable to reproduce the results when the reaction was carried out with concentrated sulfuric acid.

A second procedure involves the pyrolysis of the orthoacetate (*15, 31, 33*):

$$CH_3-C(OH)(CF_3)-CN \longrightarrow CH_3-C(OH)(CF_3)-CO_2H \longrightarrow CH_3-C(OH)(CF_3)-CO_2R$$

$$\downarrow$$

$$CH_2{=}C(CF_3)CO_2R \longleftarrow CH_3-C(O-\overset{O}{\overset{\|}{C}}-CH_3)(CF_3)-CO_2R$$

The cyanohydrin is converted to the hydroxy acid by treatment with sulfuric acid. The hydroxy acid is then esterified and acetylated. The acetate is then cracked to the desired ester. A modified procedure for the preparation of the hydroxy acid involves the conversion of the cyanohydrin to the thioamide which is in turn hydrolyzed (*31*):

$$CH_3-C(OH)(CF_3)-CN \longrightarrow CH_3-C(OH)(CF_3)-CSNH_2 \longrightarrow CH_3-C(OH)(CF_3)-CO_2H$$

A variation (*15, 31*) of this latter procedure involves the conversion of the cyanohydrin to the acetate which is then pyrolyzed to the unsaturated nitrile. The nitrile may then be hydrolyzed to the acrylic acid followed by esterification.

It is reported in a series of patents by DICKEY and COOVER (*23, 35, 36*) that either methyl-α-(trifluoromethyl)acrylate or methyl-α-(difluoromethyl)acrylate may be polymerized using either peroxide catalysts, trialkylarsines, trialkylstibines or trialkyl- or dialkyl acid phosphites. The polymers were further described as acetone soluble, moldable solids. More recently, work carried out by the General Aniline & Film Corp. (*2*) indicates that methyl-α-(trifluoromethyl)acrylate does not polymerize readily to a hgih molecular weight. Attempts to prepare homopolymers employing the following catalysts met with no success: benzoyl peroxide, 2,2′-azobis[2-methylpropionitrile], triethylphosphite or tri-*n*-butylphosphine. A substantial dose of gamma radiation ($3—4 \times 10^7$ roentgens) finally resulted in a clear transparent solid. The polymer appeared to be cross-linked but no other properties are reported.

The following tables will summarize the properties of the fluoroacrylic acids and their esters.

Table 1. *Fluorinated Acrylic Acids*

Acid	m. p. [°C]	b. p. [° C]	Ref.
$CH_2{=}CFCO_2H$	51.5—2.0		50
$CF_2{=}CHCO_2H$			50
$CF_2{=}CFCO_2H$	35.5—6.5		50
$CH_2{=}C(CF_3)CO_2H$	50—2	146—8	15

Table 2. *Esters of Fluorinated Acrylic Acids*

Ester	b. p. [° C/mm.]	n_D (t° C)	Ref.
$CH_2{=}CFCO_2CH_3$	92.5—3.5	1.3870(25)	2
	90.5—1.75/765	1.3869(20)	74, 75
$CF_2{=}CHCO_2C_2H_5$			37
$CF_2{=}CFCO_2C_2H_5$	100—0.5/750	1.3615(25)	64
$CF_2{=}CFCO_2CH_2C_3F_7$	61—1.5/50	1.3189(25)	64
$CH_2{=}C(CHF_2)CO_2CH_3$	106	1.3570(25)	2
$CH_2{=}C(CF_3)CO_2CH_3$	103.8—5.0	1.3370(20)	15 31
$CH_2{=}C(CF_3)CO_2C_2H_5$			34
$CF_3CH{=}CHCO_2C_2H_5$	112.5—5.0	1.3600(20)	69, 112
$CH_2{=}CFCO_2CH_2CH_2F$	140—4°	1.3981(20)	29

2. Synthesis of Fluorinated Esters of Acrylic Acid. The acrylates represented by the general formula $CH_2{=}CHCO_2CH_2R_f$ have received more attention than those types already discussed. The greatest portion of this work has been done by workers at the Minnesota Mining and Manufacturing Company and only a part of the work has formally been published (*11, 18, 107*). Additional information is available in the form of several Government technical reports (*6—10, 30*).

The final step in the preparation of the monomers is the same in all cases and involves the esterification of the fluoroalcohol with acrylic acid. Due to the highly acidic character of the fluoroalcohols, direct esterification is slow and impractical. Two general procedures have proven most successful for the preparation of the acrylate esters of fluoroalcohols; 1) the use of acrylyl chloride (*18*) and the alcohol either with or without the addition of an amine or an acid acceptor, 2) the use of equimolar amounts of trifluoroacetic anhydride with acrylic acid and the fluoroalcohol (*1*, *18*).

The alcohols which have been available for esterification with acrylic acid are of five general types and have been derived from three principal reactions. These types of alcohols are:
1) R_fCH_2OH, 2) $R_fOCF_2CF_2CF_2CH_2OH$, 3) $H(CF_2CF_2)CH_2OH$, 4) $R_fCHFCF_2OCH_2CH_2OH$, and 5) $R_fCH_2OCH_2CH_2OH$. The first two types can be prepared by reduction of the ester of the corresponding fluorocarboxylic acids as follows:

$$\begin{matrix} R_fCO_2H \\ \text{or} \\ R_fOCF_2CF_2CO_2H \end{matrix} \xrightarrow{H} \begin{matrix} R_fCH_2OH \\ \\ R_fOCF_2CF_2CH_2OH \end{matrix} \qquad (18)$$

The third type is prepared by the telomerization of methanol and tetrafluoroethylene:

$$CF_2{=}CF_2 + CH_3OH \longrightarrow H(CF_2CF_2)_nCH_2OH \qquad (11)$$

Table 3. *Fluorinated Esters of Acrylic Acid*

Ester	b. p. [° C/mm.]	n_D (t° C)	Ref.
$CH_2{=}CHCO_2CH_2CF_3$	45.9/125	1.3480(25)	18
$CH_2{=}CHCO_2CH_2CF_2CF_3$	50.2/100	1.3363(20)	18
$CH_2{=}CHCO_2CH_2(CF_2)_2CF_3$	51.3/50	1.3317(20)	18
$CH_2{=}CHCO_2CH_2(CF_2)_3CF_3$	57.5/30	1.3289(25)	18
$CH_2{=}CHCO_2CH_2(CF_2)_4CF_3$	63.5/20	1.3296(20)	18
$CH_2{=}CHCO_2CH_2(CF_2)_6CF_3$	55.2/2.5	1.3289(20)	18
$CH_2{=}CHCO_2CH_2(CF_2)_8CF_3$	53/34.4 × 10^{-6}	1.3279(20)	18
$CH_2{=}CHCO_2CH_2CF_2CF_2H$			5
$CH_2{=}CHCO_2CH_2(CF_2)_3CF_2H$			5
$CH_2{=}CHCO_2CH_2(CF_2)_5CF_2H$			5
$CH_2{=}CHCO_2CH_2CF_2CF_2OCF_3$			12
$CH_2{=}CHCO_2CH_2CF_2CF_2OCF_2CF_3$			12
$CH_2{=}CHCO_2CH_2CF_2CF_2O(CF_2)_2CF_3$			12
$CH_2{=}CHCO_2CH_2CF_2CF_2O(CF_2)_3CF_3$			12
CF_2—CF_2 | | $CH_2{=}CHCO_2CH_2CF_2$—CF CF_2 \ / O			12
$CH_2{=}CHCO_2CH_2CF_2CFHCF_3$			12
$CH_2{=}CHCO_2CH_2CH_2OCH_2CF_3$			12
$CH_2{=}CHCO_2CH_2CH_2OCF_2CF_2H$			12
$CH_2{=}CHCO_2(CH_2CH_2O)_2CF_2CF_2H$			12
$CH_2{=}CHCO_2CH_2CH_2OCH_2(CF_2)_2CF_3$			12
$CH_2{=}CHCO_2CH(CH_3)CF_3$	48/120	1.3573(20)	22
$CH_2{=}CHCO_2CH(CH_3)C_3F_7$	63/73	1.3400(20)	22
$CH_2{=}CHCO_2CH(C_2H_5)C_3F_7$	66/45	1.3512(20)	22
CH_3 | $CH_2{=}CCO_2CH_2CF_3$			28

The fourth type is prepared by the base catalyzed addition of ethylene glycol to a fluoroolefin:

$$R_fCF{=}CF_2 + HOCH_2CH_2OH \xrightarrow{OH^-} R_fCHFCHF_2OCH_2CH_2OH$$

The preparation of the fifth type has not been described. Table 3 indicates the acrylates reported and, where available, the physical properties.

3. Polymerization of Fluorinated Esters of Acrylic Acid. All of the acrylate monomers listed in the Table 4 will homopolymerize in either

Table 4. *Properties of Poly(1,1-dihydroperfluoroalkyl)acrylate Vulcanizates*

Monomer $CH_2{=}CHCO_2CH_2(CF_2)nCF_3$ $n =$	T_g [° C]	T_{10} [° C]	T_b [° C]	n_D^{25}
0	—10	+2	—10	1.407
1	—26	—7	—11	1.385
2	—30	—7	—15	1.367
3	—37	—7		1.360
4	—39	—7	—12 to —17	1.356
6	T_{lst}: 30—35 T_g—17	+9		1.339
8	T_{lst}: 100			

bulk, solution or emulsion recipes. Bulk polymerization with ultraviolet light as an initiator and 0.1% benzoyl peroxide as sensitizer will yield high molecular weight polymers. Reasonably high molecular weight acrylates may be obtained by solution polymerization. Hexafluoroxylene or methyl perfluorobutyrate will serve satisfactorily as solvents. Emulsion polymerization is by far the most satisfactory technique. A standard recipe is employed using potassium persulfate as initiator:

Monomer	100
Water	180
$K_2S_2O_8$	0.5
Sodium lauryl sulfate . .	3.0

Small amounts of mercaptans may be added for better control of the molecular weight and rate of polymerization. The emulsion polymerization is characterized by a sensitivity to oxygen which may cause long induction periods. The polymerization is then very rapid and may be accompanied by a rather sharp rise in temperature if precautions are not taken.

Consider first the homopolymers of the acrylates of the type $CH_2{=}CHCO_2CH_2(CF_2)_nCF_3$ (n = 0 to 8). It is observed that the polymers all exhibit low indices of refraction characteristic of fluorine-containing compounds. Bovey et al. (*11*), and Stedry et al. (*107*) report that the homopolymers of the ethyl ($n = 0$) through the hexyl ($n = 4$) derivatives are all rubbery and may be vulcanized into elastomers exhibiting good resistance to swelling in hydrocarbon solvents as well as good thermal stability. A standard silicate vulcanization recipe proved effective in curing the homopolymers.

Polymer	100
$Ca(OH)_2$	2.72
$Na_2SiO_3 \, 9H_2O$	6.72
Temp., ° F	310
Time	1 hour

The properties of the homopolymer vulcanizates prepared by the above recipe are given in the Table 4.

The properties of the acrylate vulcanizates prepared with fluorine-containing hydroxy ethers are given in Table 5.

Table 5. *Properties of Vulcanizates of Poly(fluoroalkoxyalkyl)acrylates* (*12*)

Monomer	T_g [° C]	T_{10} [° C]	n_D^{20}
A			
$CH_2{=}CHCO_2CH_2CH_2OCH_2CF_3$	—38	—3	1.419
$CH_2{=}CHCO_2CH_2CH_2OCF_2CHF_2$	—22	—9	1.412
$CH_2{=}CHCO_2(CH_2CH_2O)_2CF_2CHF_2$	—40	—15	1.422
$CH_2{=}CHCO_2CH_2CH_2OCH_2(CF_2)_2CF_3$	—45	—20	1.390
B			
$CH_2{=}CHCO_2CH_2CF_2CF_2OCF_3$	—55	—32	1.360
$CH_2{=}CHCO_2CH_2CF_2CF_2OCF_2CF_3$	—49	—23 (Silicate) Cure —34 (Oxide) Cure	1.348
$CH_2{=}CHCO_2CH_2CF_2CF_2O(CF_2)_2CF_3$	—68	—35	1.346
$CH_2{=}CHCO_2CH_2CF_2CF_2O(CF_2)_3CF_3$	—68	—31	1.346
$CH_2{=}CHCO_2CH_2CF_2{-}CF(CF_2{-}CF_2{-}CF_2{-}O{-})$ (ring: CF—CF_2—CF_2—CF_2—O)		+2	

It is interesting to note that the polymers from group A containing the —CH_2CH_2O-group do not possess improved low temperature flexibility over the 1,1-dihydroperfluoroalkyl acrylates. The short chain members of group A polymers also exhibited much higher swell in organic solvents. However, the polymers from group B, in which the ether oxygen has been moved to the three position exhibit a marked improvement in both T_g and T_{10} with no sacrafice in solvent resistance.

Polymer	100
Stearic acid	1
HAF Black	35
Sulfur	1
Triethylenetetramine	1
Cure: 60 min./300° F	

Further compounding studies have shown that for optimum vulcanizate properties and heat resistance a polyamine cure is preferred (*107*) (see preceding table). Using the above curing system the following physical properties are obtained with the homopolymers of $CH_2{=}CHCO_2CH_2(CF_2)_2CF_3$ (I) and $CH_2{=}CHCO_2CH_2CF_2CF_2OCF_3$ (II).

Copolymerization of 1,1-dihydroperfluoroalkylacrylates has been studied by BOVEY (*11*) in particular, the copolymerization of 1,1-dihydroperfluorobutylacrylate has been investigated (*98*). The fluoroacrylates copolymerize readily with butadiene, isoprene, styrene, acrylonitrile and hydrocarbon acrylates and methacrylates. The incorporation of butadiene has been shown to improve the low temperature characteristics at a sacrifice to the solvent resistance.

Table 6. *Physical Properties of Some Polyfluoroacrylate Vulcanizates*

	I	II
Modulus, 300% elongation (psi)	1020	670
Tensile (psi)	1200	1000
Ultimate elongation, %	360	400
T_{10}, °C	—7	—30
A. S. T. M. T_b, °C	—13	—40
% Swell in: 70/30 isooctane/toluene	17	15
benzene	26	19
acetone	91	64
Resilience, 25° (Bashore)	6	12

In the following table are the reactivity ratios as well as PRICE-ALFREY *Q* and *e* vaules for 1,1-dihydroperfluorobutyl acrylate in three copolymer systems. Additional copolymers of 1,1-dihydroperfluorobutyl acrylate have been described by KNOBLOCH and HAMLIN (*58*).

Table 7. *Copolymerization Characteristics for 1,1-dihydroperfluorobutyl acrylate*

M_1	r_1	r_2	e_2	Q_2
Butadiene	0.35	0.07	1.1	0.82
Methyl methacrylate	1.4	0.25	1.4	0.78
Styrene	0.33	0.07	1.1	0.64

Homopolymers of α,α,ω-trihydroperfluoroalkyl acrylates (*21*) are similar in properties to the 1,1-dihydroperfluoroalkylacrylates (*11*) and the 1-alkyl-1-hydroperfluoroalkyl acrylates (*22*). BOVEY and coworkers (*12*) have also reported this similarity but note that the terminal hydrogen atom of 1,1,5-trihydroperfluoroamyl acrylate causes a 10° lowering in the T_{10} for the corresponding 1,1-dihydroperfluoroamyl acrylate.

Homopolymerization of 1-alkyl-1-hydroperfluoroalkyl acrylates results in transparent latices and the rubbery nature of the polymer increases as the chain length increases (*22*). The homopolymer of 1-methyl-1-hydroperfluoroethyl acrylate is a tough plastic material with a softening point of 100° and exhibits no indication of crosslinking with heat. The homopolymers of 1-methyl and 1-ethyl-1-hydroperfluorobutyl acrylate are tough white elastomers.

The 1-alkyl-1-hydroperfluoroalkyl acrylates copolymerize with a variety of vinyl monomers and butadiene (*22*). Copolymerization occurs readily with styrene, acrylonitrile, vinyl acetate, maleic anhydride and benzalacetophenone. Rubbery copolymers of these acrylates with butadiene were obtained with a standard emulsion recipe.

It is interesting to note that in all cases, the curing of rubbery homopolymers of fluorine-containing acrylates is facilitated if a few hundreds percent of acrylic acid is incorporated in the polymer.

In a patent, CRAWFORD and coworkers (*28*) have reported the preparation of the homopolymer of

$$CH_2{=}\overset{\displaystyle CH_3}{\overset{|}{C}}CO_2CH_2CF_3$$

by use of benzoyl peroxide catalyst.

B. Acrylonitriles

1. Monomer Synthesis. Perfluoroacrylonitrile is prepared according to CHANEY (*16*) by the following sequence of reactions:

$$CF_2ClCF{=}CCl_2 + O_2 + Cl_2 \xrightarrow[\text{Vapor Lamp}]{\text{80-W Quartz Mercury}} \xrightarrow[NH_3]{Et_2O} CF_2ClCFCl\overset{O}{\overset{\|}{C}}NH_2$$

$$CF_2ClCFCl\overset{O}{\overset{\|}{C}}NH_2 \xrightarrow{P_2O_5} CF_2ClCFClCN \xrightarrow{Zn} CF_2{=}CFCN$$

This procedure is also applicable for the preparation of α-chloroperfluoroacrylonitrile from 1-difluoromethyl-1,2,2-trichloroethylene.

An alternate procedure for perfluoronitrile preparation is described by MILLER (*81*). A perhaloolefin, $CF_2{=}CFCClF_2$, is dissolved in either benzene, ether, pyridine or dioxane and reacted with gaseous ammonia at temperatures from 0—100° to give the nitrile directly.

Perfluoroacrylonitrile can also be prepared in a 25% overall yield by the following route (*64*):

$$CF_2{=}CFCl + ICl \longrightarrow CF_2ClCFClI$$

$$CF_2ClCFClI + CH_2{=}CH_2 \longrightarrow CF_2ClCFClCH_2CH_2I$$

$$CF_2ClCFClCH_2CH_2I \xrightarrow{OH^-} CF_2ClCFClCH{=}CH_2$$

$$CF_2ClCFClCH{=}CH_2 \xrightarrow{KMnO_4} CF_2ClCFClCO_2H$$

$$CF_2ClCFClCO_2H \longrightarrow CF_2ClCFClCO_2R \longrightarrow CF_2ClCFClCONH_2$$

$$CF_2ClCFClCOHN_2 \longrightarrow CF_2ClCFClCN \longrightarrow CF_2{=}CFCN$$

LAZERTE also reports the synthesis of perfluoroacrylonitrile from CF_3CFHCN (*63*).

α-Trifluoromethyl acrylonitrile is prepared according to DICKEY (*33*) by the following reactions:

$$CF_3\overset{\overset{\displaystyle O}{\|}}{C}CH_3 + KCN \longrightarrow CF_3-\underset{\underset{\displaystyle CN}{|}}{\overset{\overset{\displaystyle OH}{|}}{C}}-CH_3 \xrightarrow[SOCl_2]{Pyridine} \xrightarrow[150-600^\circ]{\Delta} CF_3C(CN)=CH_2$$

With this procedure, α-difluoromethyl acrylonitrile can be prepared from α,α-difluoroacetone. BUXTON (*15*) describes a similar procedure in which the cyanohydrin is converted to the acetate and pyrolyzed at 500° C to yield α-trifluoromethylacrylonitrile.

Properties of four of the fluorine-containing acrylonitriles are given in Table 8.

Table 8. *Fluorine-containing Acrylonitriles*

Compound	b. p. [° C/mm.]	n_D (t ° C)	Ref.
$CF_2=CFCN$	17.7—8.0	1.3162(10)	16
$CF_2=CClCN$	63	1.3793(24)	16
$CH_2=C(CF_3)CN$	75.9—6.2—2/759	1.3239(20)	15
$CH_2=C(CF_2H)CN$	43—8		81

2. Polymerization Reactions. Perfluoroacrylonitrile is very reluctant to undergo homopolymerization reactions. Very small quantities of polymer can be isolated when perfluoroacrylonitrile is polymerized in the presence of BF_3 or acetyl peroxide (*6—9*). A 20% conversion to an orange powder can also be obtained by using 200 megaroentgens of Co^{60} gamma radiation as initiator (*6—9, 76*).

A fairly complete study of copolymerization reactions is reported (*6—9*). These reports state that perfluoroacrylonitrile is readily hydrolyzed when emulsion polymerization is attempted. Emulsion copolymerization with either styrene or acrylonitrile results in low-conversion polymers containing 10—15% perfluoroacrylonitrile. Bulk copolymerization with styrene using cumene hydroperoxide as initiator yields a copolymer containing 20—30% fluorine, which corresponds to approximately a 1:1 copolymer (*6—9*).

Vinyl acetate copolymerizes more readily in bulk with perfluoroacrylonitrile. By varying the ratio of nitrile to vinyl acetate from 1:1 to 12:1, the fluorine content can be varied from 20—28%.

Bulk copolymers can also be obtained with butadiene and ethylene. A charge of 32 mole percent nitrile gives a butadiene copolymer containing 33 mole percent perfluoronitrile; while with ethylene a copolymer containing 20 mole percent nitrile is obtained where 50 mole percent of nitrile is charged. Vulcanizates of the perfluoroacrylonitrile-butadiene copolymer show physical properties similar to acrylonitrile-butadiene copolymers in regards to resistance to solvents (*6—9*).

Perfluoroacrylonitrile copolymerizes with 1,1-dihydroperfluorohexyl vinyl ether to give a soft plastic material containing 44 mole percent nitrile content and with vinylidine chloride, forms a white powder containing 10 mole percent nitrile content. Plastic copolymers are isolated with *i*-butyl vinyl ether and *n*-butyl vinyl ether. These polymers contain 50 mole percent nitrile and form clear flexible films which are soluble in both acetone and benzene. Isobutylene and vinyl chloride also copolymerize with perfluoroacrylonitrile (*6—9*).

Attempted copolymerizations of perfluoroacrylonitrile with 1,1-dihydroperfluorobutyl acrylate, perfluoropropene, isoperfluorobutylene, vinylidine fluoride and chlorotrifluoroethylene did not yield the desired products. Therefore, it can be seen that the reactivity of perfluoroacrylonitrile is similar to that of perfluoroolefines in that all of the materials that form copolymers lie in the low "*Q*", non-conjugated region of the Price-Alfrey diagram (*6—9*).

α-Perfluoropropyl acrylonitrile shows somewhat higher reactivity than perfluoroacrylonitrile. No homopolymer can be prepared in emulsion recipes but with styrene, a copolymer containing 52% nitrile is obtained, and with acrylonitrile, 22% α-perfluoropropyl acrylonitrile is incorporated. No hydrolysis of the monomer can be observed (*6—9*). Rubbery copolymers with butadiene are formed which contain 61% of nitrile, when mercaptan is excluded from the recipe. These copolymers show ASTM brittle points of —40 to —50° and a swelling volume of 600% in 70:30 isooctane:toluene solvent. Polymers prepared with mercaptan were over-modified. When a larger molar quantity of nitrile is introduced and the copolymer vulcanized, the vulcanizate was too leathery and hard with a brittle point above 0° C and a swelling volume of 75% (*6—9*). Dickey (*33*) reports that α-trifluoromethyl- and α-difluoromethyl acrylonitrile homopolymerize, and copolymerize with styrene and butadiene. The 2-trifluoromethyl-derivative also is claimed to copolymerize with acrylonitrile, methacrylonitrile, acrylamide, and dimethylfumarate and vinylidine chloride. These nitriles can also be homopolymerized using trimethyl arsine (*23*) and trialkyl- or dialkyl acid phosphites (*36*).

Workers at the M. W. Kellogg Co. (*25*) report the investigation of α-trifluoromethyl acrylonitrile as a comonomer with various dienes and they conclude that the resulting copolymers have poorer low temperature flexibility than non-fluorinated acrylonitrile copolymers.

α-Trifluoroacetoxy acrylonitrile can be homopolymerized and copolymerized with vinyl chloride and styrene. α-Difluoroacetoxy acrylonitrile can be copolymerized with methyl methacrylate. The polymers are claimed to have higher melting points and greater heat resistance than non-halogenated acetoxyacrylonitriles (*38*).

C. Acrylamides

1. Monomer Synthesis. Several fluoroalkylacrylamides and methacrylamides were prepared and their polymerization studied. HALPERN and coworkers (*47*) employed the following general procedure for the preparation of the monomers listed in Table 9:

$$R_fCO_2H + CH_3OH \longrightarrow R_fCO_2CH_3$$

$$R_fCO_2CH_3 + RNH_2 \longrightarrow R_fCONHR$$

$$R_fCONHR + LiAlH_4 \longrightarrow R_fCH_2NHR$$

$$R_fCH_2NHR + CH_2{=}C(R')COCl \longrightarrow CH_2{=}C(R')CON(R)(R_f)$$

The above procedure is straight-forward and gives good overall yields of acrylamides. A word of caution should be interjected at this point concerning the reduction of the fluorine-containing amides with lithium aluminium hydride to give the fluoroamines. Several explosions have been reported to occur when reductions of this type are carried out and all possible precautions should be taken during this step.

Table 9. *Fluoroacrylamides* (*47*)

Type	R—	m. p. [° C]	b. p. [° C/mm.]	n_D [t ° C]
$CH_2{=}CHCONHR$	CF_3CH_2—	74.5—75		
	n-$C_3F_7CH_2$—	57.4—57.6		
$CH_2{=}CHCON(CH_2C_3F_7)(R)$	CH_3—		79/11	
	C_2H_5—		95/15	1.3829(20)
	n-C_4H_9—		112/15	1.3930(20)
	iso-C_4H_9—		107/15	1.3901(20)
$CH_2{=}C(CH_3)CON(CH_2C_3F_7)(R)$	CH_3—		77/8	1.3763(20)
	C_2H_5—		107/29	1.3844(25)
	n-C_4H_9—		114/11	1.3906(20)
	iso-C_4H_9—		90/5	1.3947(20)

The procedure by which the reaction mixture is worked up will also determine the yield of fluoroamine obtained. The procedure initially employed called for the decomposition of any excess lithium aluminium hydride with ethyl acetate followed by treatment with cold dilute sulfuric acid. The bisulfate salt thus formed was washed and dried and the amine regenerated by treatment with a 50% sodium hydroxide solution. Later experiments disclosed that improved yields of amines could be obtained if the hydrolysis were carried out on the basic side after which the amine could be isolated directly from the ether layer.

2. Polymerization Studies. The first six fluoroacrylamides listed in Table 9 were found to homopolymerize rapidly to high conversion in either bulk, solution, or emulsion systems (*57*). As with the fluoroacrylates, long induction periods were encountered if precautions *were not taken* to exclude all oxygen. Bulk polymerizations were initiated with benzoyl peroxide and solution polymerizations were carried out in benzene with the same catalyst. Emulsion polymerizations were carried out using potassium persulfate at 50°. It was observed that after the induction period was passed polymerization was extremely rapid and was substantially complete after 10 minutes at 50°.

All the homopolymers of this series were thermoplastic solids which could be cast into transparent films or drawn from a melt or solution into weak fibers. The acrylamides prepared from the primary amines softened about 140° and those from the secondary amines softened about 100°. The polymers appeared to exhibit a reasonable degree of thermal stability after melting to viscous liquids at about 200°. These homopolymers also exhibited a high degree of elasticity when plasticized with common solvents such as methanol and benzene.

The fluoroacrylamides copolymerize readily with dienes, vinyl ethers, alkyl- and fluoroalkyl acrylates. Two copolymers exhibited unusually high thermal stability. The first a stiff waxy copolymer of $CH_2{=}CHCONHCH_2CF_3$ and $CH_2{=}CHCO_2CH_2C_3F_7$ exhibited only slight coloring upon heating to 330°. The second copolomer of

$$CH_2{=}CHCON\begin{matrix} \diagup CH_2C_3F_7 \\ \diagdown C_4H_9 \end{matrix} \quad \text{and} \quad CH_2{=}CHCO_2CH_2CF_3$$

lost only 5.4% weight after heating for 72 hours at 178°.

D. Vinyl Ethers

1. Monomer Synthesis. Although many fluorine-containing vinyl ethers are known, very little published information is available on the preparation of those ethers whose polymerization reactions are known.

CORLEY (*27*) describes the preparation of 1-methoxy-1-2-difluoro-2-chloroethylene by the base catalyzed addition of methanol to 1,1,2-trifluoro-2-chloroethylene and subsequent dehydrofluorination. Two cyclic vinyl ethers, 1-ethoxy perfluorocyclobutene-1 and 1,2-diethoxy perfluorocyclobutene-1, are known to polymerize. The diethoxy compound can be prepared by reacting perfluorocyclobutene with ethanol in the presence of potassium hydroxide (*82*). If benzyltrimethylammonium hydroxide is used as catalyst, the monoethoxy derivative is the major product (*4*).

The preparatory procedures for 1,1-dihydroperfluoroalkyl vinyl ethers and β-hydroperfluoroalkyl vinyl ethers are unpublished.

2. Polymerization Reactions. Polymerization reactions of two types of vinyl ethers, β-hydroperfluoroalkyl and 1,1-dihydroperfluoroalkyl, are reported by Bovey (*6—9*). These ethers exhibit interesting polymerization characteristics. Whereas the hydrocarbon vinyl ethers polymerize very reluctantly in free radical systems, the β-hydroperfluoroalkyl vinyl ethers polymerize very rapidly under these conditions but do not polymerize with ionic or acid catalysis. The 1,1-dihydroperfluoroalkyl vinyl ethers homopolymerize with either free radical or acid catalysis.

Beta-hydroperfluoroethyl vinyl ethers forms homopolymers with peroxide catalysts in bulk or emulsion. These polymers range from resinous to soft materials; swell in aromatic solvents; exhibit a Gehman T_{10} of +20.5° and have a intrinsic viscosity of 1.0. Flexible, slightly elastic homopolymers of β-hydroperfluoropropyl vinyl ether are formed in emulsion systems exhibit less swelling in aromatic solvents than the ethyl derivative and have a Gehman T_{10} of +15.5° (*6—9*).

1,1-Dihydroperfluorobutyl vinyl ether gives a soft rubbery homopolymer with BF_3 at low temperatures and a tough resinous product in low yield with benzoyl peroxide, at high temperatures. This vinyl ether forms copolymers with 1,1-dihydroperfluorobutyl acrylate and acrylonitrile but not with styrene. Homopolymers of 1,1-dihydroperfluorohexyl vinyl ether can also be prepared. This ether is more reactive than the butyl derivative but exhibits the same copolymerization reactions (*6—9*).

Perfluorobutadiene copolymerizes with 1,1-dihydroperfluoroethyl vinyl ether both in bulk and in emulsion to give polymers which are rubbery even at relatively high conversions. These products, when properly stabilized, show weight loss of only about 8% and only slight stiffening after 500 hours at 200° and they are resistant to ozone. Study of these materials at elevated temperatures indicates that while HF is liberated together with other products, the chief loss in weight is probably due to the splitting out of the side-chain of the vinyl ether. Despite this, the copolymers containing higher proportions of vinyl ether appear to be more stable at high temperature (*6—9*). Testing of a vulcanizate of a perfluorobutadiene-1,1-dihydroperfluorobutyl vinyl ether copolymer indicates good resistance to solvents and dry heat. Unfortunately, the results are hard to duplicate and the products difficult to analyze. At this time, no work is in progress on these systems.

Workers at the M. W. Kellogg Co. (*26*) report that β-hydroperfluoroethyl vinyl ether forms resinous or rubbery copolymers in fair yields with dienes such as fluoroprene and 2-trifluoromethylbutadiene. Only small amounts of the ether are incorporated and consequently, fuel resistance is uneffected. When β-hydro-β-chloroperfluoroethyl vinyl ether is incorporated into a terpolymer with 1,1,2- and 1,1,3-trifluorobutadiene, the Gehman T_5 is lowered by 6°.

The polymerization of two cyclic vinyl ethers, 1-ethoxy perfluorocyclobutene-1 and 1,2-diethoxyperfluorocyclobutene-1, is now being investigated (*55*). Preliminary results show that these cyclic vinyl ethers form homopolymers; copolymers with the diethoxy compound exhibit the best reactivity.

Vinyl ethers containing fluorine on the vinyl group are generally unreactive. In the case of $CFCl{=}CFOCH_3$, a small amount of unstable homopolymer can be isolated using $SnCl_4$ as catalyst (*27*).

E. Vinyl Esters

Several vinyl esters of perfluorocarboxylic acids are known and these monomers exhibit polymerization reactivities similar to that of their hydrocarbon analogs. Unfortunately, both the monomers and homopolymers, as a group, are hydrolyticly unstable thus limiting their utility. Several references appear in the literature, but most of the pertinent information is contained in one patent (*51*) and two recent articles (*46, 93*).

1. Monomer Synthesis. The most useful synthesis of vinyl esters is the mercury-catalyzed addition of perfluoro mono-carboxylic acids to acetylene. HOWK (*51*) describes the synthesis of vinyl trifluoroacetate by the liquid phase addition of trifluoroacetic acid to acetylene in the presence of mercuric oxide, mercuric sulfate and hydroquinone. DICKEY (*40*) reports the same reaction using magnesium oxide and oleum. The vinyl esters can also be prepared according to COOVER (*24*) by vapor-phase addition using either zinc oxide or cadmium oxide as catalyst.

The most satisfactory procedure is described by REID and coworkers (*93*). The mercuric salt of the perfluoro monocarboxylic acid is prepared from red mercuric oxide and the perfluoro monocarboxylic acid. The anhydride of the perfluoro acid is added to the reaction mixture to absorb water produced, and this mixture is reacted with acetylene. In cases where the anhydride is not available, trifluoroacetic anhydride can be substituted. Examples of some preparative reactions are shown below:

(1) $C_2F_5CO_2H + CH{\equiv}CH + HgO + (C_2F_5CO)_2O \xrightarrow{61\%} C_2F_5CO_2CH{=}CH_2$

(2) $C_4F_9CO_2H + CH{\equiv}CH + HgO + (C_4F_9CO)_2O \xrightarrow{73\%} C_4F_9CO_2CH{=}CH_2$

(3) $C_9F_{19}CO_2H + CH{\equiv}CH + HgO + (C_2F_5CO)_2O \xrightarrow{38\%} C_9F_{19}CO_2CH{=}CH_2$

Properties of the vinyl esters are given in Table 10.

Table 10. *Vinyl Esters* — $R_fCO_2CH{=}CH_2$

Rf—	b. p. [° C/mm.]	n_D [t ° C]	Reference
CF_3—	39/747	1.3151(25)	51
	39.5—40.5	1.3106(25)	
C_2F_5—	58/745	1.3095(20)	93
C_3F_7—	78—79/745	1.3086(20)	93
C_4F_9—	97—99/732	1.3116(20)	93
C_5F_{11}—	65—66/100	1.3115(25)	51
Cyclo-C_6F_{11}—	59/45	1.3362(20)	93

2. Polymerization Reactions. Homopolymerization of the vinyl esters of perfluoroacids can be accomplished using either peroxide or ultraviolet light catalysis (*93*).

Table 11. *Homopolymers of Vinyl Esters of Perfluorocarboxylic Acids* (*93*)

$(-CH_2-CH(CO_2Rf)-)_n$			Polymer	
Rf	n_D^{25}	T_g [° C]	Appearance	Nature
CF_3—	1.375	above 75	clear, tough	hard, sl. flex.
C_2F_5—	1.3642	42	clear, tough	hard, sl. flex.
C_3F_7—	1.3630	27	clear, flex.	
C_4F_9—	1.3530	15—20	clear, flex.	
C_5F_{11}—	1.3440	—9	clear, flex., sl. elastic	
Cyclo-C_6F_{11}—	1.3691	54	clear, hard, flex.	

The glass temperatures of the resultant polymers range from above 75° for vinyl perfluoroacetate to —20° for vinyl perfluorocaprate. Bulk polymerization of vinyl perfluoroacetate gives a hard fibrous polymer wich softens at 172° — much higher than polyvinyl acetate (105°). Methyl perfluorobutyrate is preferred for solution polymerization and yields a powdery product with lower $\overline{DP}$ than bulk techniques. Polyvinyl perfluoroacetate is insoluble in the monomer and polymerization leads to no increase in $\overline{DP}$ with increase in conversion. It is believed that substitution of the CF_3-group for CH_3-reduces branching and chaim transfer (*93*). The homopolymers form clear films; however, discoloration occurs on standing due to hydrolysis. Polyvinyl trifluoroacetate is readily orientated by stretching and can be hydrolyzed by gaseous ammonia to highly birefringent polyvinyl alcohol.

The vinyl esters copolymerize readily with vinyl acetate and methyl methacrylate and to a lesser extent with styrene, methyl acrylate and acrylonitrile. Bulk and solution techniques are the most satisfactory for copolymer preparation (*93*). HAAS and coworkers report (*46*), that contrary to the previous literature (*51*), aqueous suspension polymerizations were unsuccessful due to monomer hydrolysis. Reactivity ratio determinations of vinyl perfluoroacetate (M_1) and vinyl acetate (M_2) give values of $r_1 = 0.32$ and $r_2 = 0.6$, indicating that each radical prefers to add the other monomer. Values of $Q = 0.022$ and $e = 0.82$ for vinyl perfluoroacetate are found by using average values of $Q = 0.024$, $e = -0.47$ for vinyl acetate.

F. Unsaturated Ketones

A series of three perfluoroalkyl propenyl ketones are the only fluorine-containing α,β-unsaturated ketones whose polymerization reactions are known (*91*). Although these ketones do not homopolymerize, they readily form copolymers with a wide variety of vinyl monomers and dienes. Much work remains to be done in the determination of the effects of fluorine substitution on the alkyl group and on the α- and β-positions of the double bond.

1. Monomer Synthesis. The perfluoroalkyl propenyl ketones are prepared by reacting a perfluorocarboxylic acid with allylmagnesium bromide and subsequent rearrangment according to the reaction sequence below (*91*):

$$(1)\quad CH_2{=}CHCH_2MgBr + R_fCO_2H \longrightarrow CH_2{=}CHCH_2\overset{\overset{\displaystyle O}{\|}}{C}R_f$$

$$(2)\quad CH_2{=}CHCH_2\overset{\overset{\displaystyle O}{\|}}{C}R_f \longrightarrow CH_3CH{=}CH\overset{\overset{\displaystyle O}{\|}}{C}R_f$$

The β,γ-unsaturated ketone isomerizes to the α,β-unsaturated ketone on standing, and pure propenyl ketone is isolated in 40—50% yields by fractional distillation. Physical constants of the ketones are given in Table 12.

Table 12 (*91*). *Fluorine-containing Propenyl Ketones* — $R_f\overset{\overset{\displaystyle O}{\|}}{C}CH{=}CHCH_3$

R_f—	Boiling Point b. p. [° C/mm.]	n_D (t ° C)
CF_3—	85—6/757	1.3583(20)
C_3F_7—	116.7/745	1.3400(20)
C_7F_{15}—	55.0—5.5/5	1.3340(20)

2. Polymerization Reactions. The perfluoroalkyl propenyl ketones do not homopolymerize using conventional free-radical techniques. They do, however, undergo radical catalyzed copolymerization with a variety of vinyl monomers such as acrylonitrile, ethyl acrylate, styrene, vinyl acetate, 1,1-dihydroperfluorobutyl acrylate and *m*-trifluoromethyl styrene (*91*). In comparing the reactivity of the three ketones, perfluoropropyl propenyl ketone exhibits the best balance of properties. Data on the bulk copolymers of the propenyl ketones is shown in Table 13.

Table 13.
Bulk Polymerizations of Perfluoroalkyl Propenyl Ketones with Vinyl Monomers

Comonomer	Polm. time, hrs.	Conversion %	Inherent Viscosity (benzene)	Softening Point [° C]	Change Ratio Ketone per Comonomer	Fluorine %	Ketone Incorp. [Wt. %]	Appearance of Polymer
				Perfluoromethyl Propenyl Ketone				
Acrylonitrile	7	35	0.8[1]	275	40/60	Trace	5[2]	Powder
Ethyl Acrylate	7	25			30/70	3.7	9	Gum
Vinyl Acetate	48	25	0.9	90—95	40/60	23.6	57	Powder
Styrene	7	55	0.5	145—150	40/60	14.0	34	Powder
				Perfluoropropyl Propenyl Ketone				
Acrylonitrile	7	40	0.85[1]	275	40/60	Trace	5[2]	Powder
Ethyl Acrylate	7	25			30/70	7.5	13	Gum
Vinyl Acetate	48	40	0.57	83—88	40/60	36.8	65	Powder
Styrene	7	50	0.30	140—145	40/60	27.5	50	Powder
				Perfluoroheptyl Propenyl Ketone				
Acrylonitrile	7	45	1.7[1]	275	40/60	4.4	7	Powder
Ethyl Acrylate	7	25			30/70	15.5	24	Gum
Vinyl Acetate	48	40	0.7	75—80	40/60	49.4	75	Powder
Styrene	7	50	0.3	130—135	40/60	33.6	52	Powder

Determination of the reactivity ratio of perfluoroalkyl propenyl ketone with styrene is now in progress and preliminary results indicate the ketone (M_2) has a $r_2 = 0$ and r_1 ranges from 0.5—1.0 (*20*).

Copolymers of the fluorine-containing ketones with dienes such as butadiene, isoprene, 1,1,2-trifluorobutadiene and 1,1,2-trifluoro-3-chlorobutadiene can be prepared; however, copolymerization does not take place with 1,1,2-trifluoro-3,4,4-trichlorobutadiene or 1,1,2-trifluoro-3,4-dichlorobutadiene using a variety of catalysts. Characterization and evaluation of the ketone-diene copolymers is now being completed (*20*).

III. Condensation Polymers

A. Polyethers

Research was initiated on the polyethers with the expectation that the introduction of the —C—O—C— links would improve the elasticity without sacrificing stability and organic solvent resistance of fluorocarbon

[1] Inherent viscosity determined in N,N'-dimethylformamide.
[2] Estimated from infrared spectrum.

Polymers. This expectation has not been realized since only low molecular weight polymers are obtained. The liquid homopolymers do show promise as low-load lubricants.

1. Monomer Synthesis. Several fluorine-containing epoxides are known, however only those epoxides whose polymerization reactions are known are listed in Table 14. The procedure that McBEE and BURTON (*66*) report for the conversion of 1,1,1-trifluoroacetone to 3,3,3-trifluoro-1,2-epoxypropane can be applied successfully in the preparation of all the epoxides. This method is illustrated below.

Table 14. *Some Fluorine-containing Epoxides*

Compound	b. p. [° C/mm.]	n_D (t ° C)	Ref.
$CF_3\overset{O}{\overbrace{CHCH_2}}$	39.1—9.3/748	1.2997(20)	105
$CF_3\overset{O}{\overbrace{C(CH_3)CH_2}}$	56.0/756	1.3125(25)	105
$CF_3\overset{O}{\overbrace{CHCH}}CH_3$	58.5—9.0/747	1.3167(20)	92
$CF_3\overset{O}{\overbrace{CHCH}}C_2H_5$	78.8—9.0/745	1.3340(20)	92
$CF_3\overset{O}{\overbrace{CHC}}(CH_3)_2$	71.3—1.9/747	1.3292(20)	92
$C_3F_7\overset{O}{\overbrace{CHCH}}CH_3$	93.5/748	1.3091(20)	92
$C_3F_7\overset{O}{\overbrace{CHCH}}C_2H_5$	110.5—1.0/749	1.3218(20)	92
$C_3F_7\overset{O}{\overbrace{CHC}}(CH_3)_2$	102.5—3.0/747	1.3187(20)	92

(1) $$CF_3\overset{\overset{O}{\|}}{C}CH_3 \xrightarrow[Br.]{H_2SO_4} CF_3\overset{\overset{O}{\|}}{C}CH_2Br$$

(2) $$CF_3\overset{\overset{O}{\|}}{C}CH_2Br + LiAlH_4 \longrightarrow CH_3CH(OH)CH_2Br$$

(3) $$CF_3CH(OH)CH_2Br + 50\%\ \text{aq. NaOH} \longrightarrow CF_3\overline{CHCH_2O}$$

The alkyl perfluoroalkyl ketones are prepared by reacting one mole of perfluorocarboxylic acid with three moles of alkyl Grignard reagent according to the procedure of DISHART and LEVINE (*41*). Good yields of the ketones also are obtained by the reaction of 1.1 moles of alkyl Grignard reagent with one mole of the lithium salt of the perfluorocarboxylic acid (*91*).

Several alternative procedures for the preparation of fluorine-containing epoxides are described by SMITH, MURCH and PIERCE (*105*), however, only one yielded the desired product.

2. Polymerization Reactions. Polymerization studies of the 1,2-epoxypropanes by SMITH, MURCH and PIERCE (*105*) indicate that aluminium chloride and ferric chloride are the most efficient catalysts. Aluminium chloride catalysis results in rapid polymerization and low molecular weight liquid products. The rate of polymerization is slower with ferric chloride and its use with 3,3,3-trifluoro-1,2-epoxypropane results in a solid polyether with a weight-average molecular weight of 230000. Optimum conditions for polymerization are 2—3 weight percent of catalyst at 90—100° for sixty-four hours.

Copolymers of 3,3,3-trifluoro-1,2-epoxypropane with 3,3,3-trifluoro-2-methyl-1,2-epoxypropane can be prepared and they range from liquids to waxy solids.

Attempts to convert the liquid polymers to solids by extending with diisocyanates and anhydrides have not been successful. The solid polymers exhibit good thermal stability and solvent resistance but vulcanization studies are inconclusive. The liquid homopolymers do show promise as low load lubricants.

Ferric chloride and aluminium chloride are also the most useful catalysts for the polymerization of the six 2,3-epoxybutanes listed in Table 14. All of the polymeric materials isolated in these experiments are liquids of relatively low molecular weight. The tendency of the 2,3-epoxybutane to polymerize decreases as the amount and complexity of substition on the ethylene oxide nucleus increases and all of these monomers are less reactive towards polymerization than 3,3,3-trifluoro-1,2-epoxypropane (*56*).

In addition to aluminum chloride and ferric chloride, BF_3 in the form of the etherate can be used to polymerize the fluorine-containing epoxides. Potassium hydroxide and a ferric chloride-propylene oxide complex can also be used as catalysts. Free-radical catalysis, gamma radiation and a heterogeneous catalyst prepared from triethyl aluminum and titanium tetrachloride do not induce polymerization (*56*).

B. Polyesters

Research in the area of fluorine-containing condensation polymers has not been reported until very recently and indications are that a high degree of thermal stability is obtained from this type of system.

In considering polyesters which contain fluorine, two general types have been investigated. These are based on the following reactants; 1) a fluorocarbon dibasic acid and a hydrocarbon diol and 2) a hydrocarbon dibasic acid and a fluorocarbon diol. A brief review of the chemistry of the preparation and the reactions of the fluorine-containing dicarboxylic acid and diols will serve to explain the choice of the most promising system and to anticipate the problems to be encountered with each system.

1. Monomer Synthesis. The most practical synthesis of the perfluorodicarboxylic acids is that of McBEE (*73*):

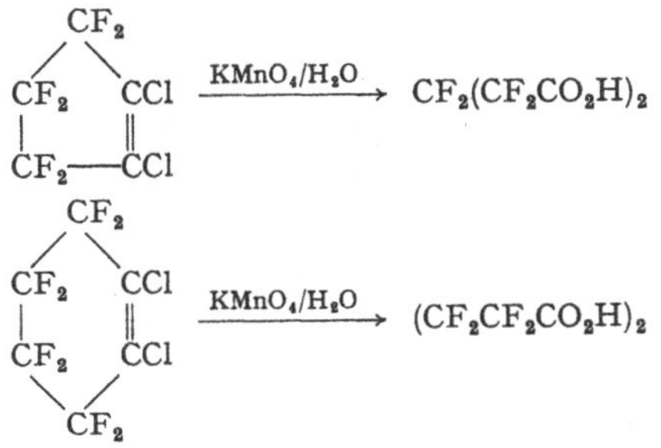

$$\text{cyclo-}[CF_2-CF_2-CF_2-CCl{=}CCl-] \xrightarrow{KMnO_4/H_2O} CF_2(CF_2CO_2H)_2$$

$$\text{cyclo-}[CF_2-CF_2-CF_2-CF_2-CCl{=}CCl-] \xrightarrow{KMnO_4/H_2O} (CF_2CF_2CO_2H)_2$$

The fact that the perfluorocarboxylic acids have a tendency to form stable hydrates requires special attention during purification. Usually the acid may be obtained from the hydrate by allowing it to remain in a desiccator over phosphoric anhydride at ambient or slightly elevated temperatures.

Diacyl chlorides of fluorocarbon dibasic acids are difficult to prepare in a high state of purity. Perfluoroadipyl chloride may be prepared by refluxing a mixture of the acid and thionyl chloride. If perfluoroadipic acid dihydrate, is used only the acid anhydride is obtained. Similarly, the reaction between perfluoroglutaric acid and thionyl chloride yields the anhydride. Phosphorus pentachloride may be used, but separation of the diacyl halide from the resulting phosphorus oxychloride is difficult (*99*).

Preparation of diethyl esters of perfluorodicarboxylic acids presents no special problems. Good yields of diethyl esters may be obtained by refluxing the acid or its hydrate with an excess of ethanol and a small amount of sodium bisulfite in benzene. The water and excess benzene are azeotroped from the mixture and the crude ester, dreid and distilled.

The preparation of fluorine-containing diols involves reduction of the diethyl ester of the acid as follows (*67*).

$$CF_2(CF_2CO_2C_2H_5)_2 + LiAlH_4/(C_2H_5)_2O \longrightarrow CF_2(CF_2CH_2OH)_2$$

2. Polyesterification. Let us first consider the polyester prepared from a hydrocarbon diol and a perfluorodicarboxylic acid. One would expect that a polyester of this type would be readily hydrolyzed in

Table 15. *Reactions of Dicarboxylic Acids and Fluorine-containing Diols (99)*

Starting materials (equimolar quantities)	Catalyst [wt. % $ZnCl_2$]	Reaction temp. [° C]	Reaction time [hrs.]	Neutral equiv.	Physical form
$(CH_2)_4(CO_2H)_2 + HOCH_2(CF_2)_3CH_2OH$	0.01	150, 180, 200, 215	240	9000	Brown, viscous liquid; crystallized very slowly
$(CF_2)_4(CO_2H)_2 + HOCH_2(CF_2)_3CH_2OH$	0.01	150, 180, 200, 215	240	4000	Amber, viscous liquid; crystallized very slowly
$(CF_2)_3(CO_2H)_2 + HOCH_2(CF_2)_3CH_2OH$	0.05	150, 180, 200	140	10000	Tan, semisolid
$(CH_2SCH_2CO_2H)_2 + HOCH_2(CF_2)_3CH_2OH$	0.15	150, 180	162	20000	Brown, slightly rubberlike; brittle —40°
$CH_2(SCH_2CO_2H)_2 + HOCH_2(CF_2)_3CH_2OH$	0.15	150, 180	162	6000	Light brown, waxy solid; m. p. 55°
$(CF_2)_3(CO_2H)_2 + HOCH_2(CF_2)_4CH_2OH$	0.05	150, 180, 200	140	6000	Light brown, semisolid
$(CH_2SCH_2CO_2H)_2 + HOCH_2(CF_2)_4CH_2OH$	0.05	150, 180	193	9000	Amber, slightly rubberlike; brittle —38°
$CH_2(SCH_2CO_2H)_2 + HOCH_2(CF_2)_4CH_2OH$	0.05	150, 180	193	6000	Tan, waxy solid; m. p. 70—75°

Table 16. *Reactions of Dicarboxylic Acid Chlorides and Fluorine-containing Diols (99)*

Starting materials (equimolar quantities)	Reaction time [hrs.]	Viscosity (poises)	M_n	x_n	Z_w	Physical form
$(CH_2)_4(COCl)_2 + HOCH_2(CF_2)_3CH_2OH$	9	1370 at 205°	17400	108	1400	Pale tan, rubberlike; crystallized very slowly; brittle —65°; m. p. 30°
$(CF_2)_4(COCl)_2 + HOCH_2(CF_2)_3CH_2OH$	21.5	25.3 at 110°	6570	28.2	362	Light tan, viscous liquid; crystallized very slowly; brittle —55°; m. p. 35°
$(CH_2)_3(COCl)_2 + HOCH_2(CF_2)_3CH_2OH$	12	60.7 at 110°	6020	39.1	463	Tan, waxy solid, m. p. 40° brittle —50° to —55°
$C_3F_7CH(CH_2COCl)_2 + HOCH_2(CF_2)_3CH_2OH$	22	33.0 at 110°	7860	33.0	392	Tan, viscous liquid; brittle —25 to —30°
$(CH_2)_4(COCl)_2 + HOCH_2(CF_2)_4CH_2OH$	5	1200 at 205°	18300	98.6	1372	White, waxy solid; m. p. 60—70°
$(CF_2)_4(COCl)_2 + HOCH_2(CF_2)_4CH_2OH$	13	11.3 at 110°	5260	20.4	281	Tan, waxy solid; m. p. 65—70°
$(CH_2)_3(COCl)_2 + HOCH_2(CF_2)_4CH_2OH$	13	17.2 at 110°	6600	25.1	322	Tan, waxy solid; m. p. 40°; brittle —50 to —55°
$C_3H_7CH(CH_2COCl)_2 + HOCH_2(CF_2)_4CH_2OH$	8	45.0 at 110°	8770	33.3	427	Tan, viscous liquid; brittle —25°

dilute base. The polyesters of the second type, those prepared from fluorinated diols and hydrocarbon diacids offer the advantage that the hydrolytic stability of the ester linkage is improved. The presence of fluorine on the beta carbon atom of the diol would preclude decomposition by olefin formation.

SCHWEIKER and coworkers (*99, 101, 102*) have contributed most of the information concerning the preparation of fluoropolyesters.

Considering first the case of direct esterification of either a perfluorodicarboxylic acid or its diacyl chloride with various diols, it is observed that only moderate molecular weight polymers are obtained (*100*). The reaction of perfluoroglutaric acid and 1,5-pentanediol with 0.05 weight percent of $ZnCl_2$ at temperatures of 150 to 180° for 199 hours gives a black viscous liquid. Molecular weight data could not be obtained from determination of neutralization equivalent, due to the dark color. When the same conditions and catalysts are employed with 3,5-dithiaheptanediol and 3,6-dithiaoctanediol, extensive decomposition was observed with the resultant formation of dithian. The addition of 0.05% of triphenylphosphite failed to prevent discoloration.

The second general case, involving the polyesterification of fluorine-containing diols with hydrocarbon dicarboxylic acids has received much attention. Three general methods for polymerization are: 1) direct esterification using zinc chloride as a catalyst, 2) transesterification; and 3) reaction of the diol with a diacyl chloride. The first two methods are slow reactions, producing only low molecular weight polymers. Table 15 gives the results of a number of direct esterification reactions (*99*).

In contrast the reaction of di-acid chlorides and fluorine-containing diols is shown in Table 16 (*99*).

It is readily apparent from the Table 16 that the highest degree of polymerization (X_n) was attained with adipyl chloride. The polyester

Table 17 (*88*). *Properties of Polyhexafluoropentylene Adipate Rubber*

Physical Properties	Original	70 hrs. — 350° F. Diester-base oil	168 hrs. — 400° F. Air
Tensile strength, psi	2200—2400	1000—1330	940—1400
Elongation, %	125—225	150—175	150—175
Hardness, Shore A	70—80	64—69	75—78
Set at break, %	0—1	6—7	18—37
Weight loss, %		6	
Brittle point	—98° F		
Freezing point	—71° F		
Gehman stiffness, T_{10}[1]	—62° F		
Temperature retraction, TR-10	—39° F		

[1] ASTM D 1053—54 T.

prepared from adipyl chloride and 1,1,5,5-tetrohydrohexafluoropentane-1,4-diol can be cured with dicumyl peroxide after compounding with carbon black and calcium carbonate to give rubbery material with the following physical properties (see table 17).

The vulcanizate exhibits a low brittle temperature. However, crystallization is a problem if the polymer is kept at low temperatures for extended periods of time. The incorporation of approximately 10% of isophthalic acid into the polymerization mixture tends to overcome this deficiency (*102*).

C. Polysiloxanes

Polysiloxane elastomers are unique because they are serviceable over the temperature range of —65 to 260°. One important deficiency is their poor resistance to organic solvents, aircraft fuels, synthetic lubricants and hydraulic fluids. In the search for means of overcoming this deficiency, it was recognized that fluorine on aryl or alkyl groups attached to the silicon atom would be a logical approach. By the same token, polysiloxane fluids have an extremely wide liquid temperature range and excellent viscosity temperature relationships. But these fluids are deficient in their ability to lubricate without excessive wear under high loads. Again, it was though that fluorine could overcome this deficiency when incorporated into the monomer unit.

1. Monomer Synthesis. There are few references to the preparation of nuclear fluorinated aromatic silicone monomers. In general, the method of synthesis follows the procedure outlined below:

$$2\,F\text{-}C_6H_4\text{-}MgBr + SiCl_4 \longrightarrow 2\,(F\text{-}C_6H_4)_2SiCl_2$$

Variations were also accomplished by using either methyltriethoxysilane or dimethyldiethoxysilane with the Grignard reagent (*113*).

The preparation of polymerizable trifluoromethylphenylsilanes has also been reported. Monomeric dichloro — and diethoxy silanes containing the trifluoromethylphenyl group were prepared from the Grignard reagent:

$$2\,CF_3C_6H_4\text{—}MgBr + SiCl_4 \longrightarrow (CF_3C_6H_4)_2\text{—}SiCl_2 \quad (45,\ 62)$$

Such other polymerizable silanes as bis(3-trifluoromethylphenyl)dichlorosilane (*45, 84*), (3-trifluoromethylphenyl)trichlosilane (*45, 62*) and [bis(trifluoromethylphenyl)]dichlorosilane were likewise prepared by this procedure (*59*). Fluorination of trichloromethylphenyl silane by the reaction involving $SbF_3 + SbCl_5$ is another route to monomer synthesis (*45*).

In contrast to fluorine-containing arylsilanes, a considerable amount of work has been done on fluoroalkylsilane synthesis and five general methods have been reported for their preparation: a) the reaction of perfluoroalkyl halides with silicon; b) the addition of chlorosilanes to fluoroolefins; c) the addition of fluoroalkyl halides to vinyl silanes; d) the dimerization of fluoroolefins and allyl silanes and e) the reaction of fluoroalkyl organometallic compounds with chlorosilanes or alkoxysilanes.

a) The direct method involving the reaction of CF_3Cl with silicon-copper alloy attempted at temperatures below 400° was fruitless; however at 500—1000° an exothermic reaction occured yielding only chlorofluorosilanes (*53*).

Patent claims were made for the preparation of CF_3SiBrF_2 and CF_3SiF_3 (*104*) and $(CF_3)_2SiX_2$ (*83*) by this process at 500° (*104*) and at 900 to 1200° (*83*) but no data concerning yields a physical properties of these products were given. Considerable doubt remains concerning the efficacy of this method of fluoroalkyl silane preparation.

b) The addition of silanes to fluoroolefins is a variation of the reaction first reported by SOMMER, PIETRUSZA, and WHITMORE (*106*) who added trichlorosilane to octene-1:

$$Cl_3SiH + C_6H_{13}CH{=}CH_2 \xrightarrow{\text{peroxide}} Cl_3SiCH_2CH_2C_6H_{13}$$

WAGNER (*110*) effected the addition of trichlorosilane to vinylidine fluoride with the use of a platinum-on-charcoal catalyst; the reaction was carried out in an autoclave under 30 to 1000 psi pressure at 100 to 300° and $Cl_3SiCH_2CHF_2$ was obtained. HASZELDINE and MARKLOW (*48*) reported that the thermally catalyzed addition of trichlorosilane to tetrafluoroethylene resulted in a series of telomers of the type, $H(CF_2CF_2)_nSiCl_3$, where $n = 1$ to 10. Additions of methyl dichlorosilane to olefins of the type $R_fCH = CH_2$ (where $R = CF_3$, C_2F_5 and C_3F_7) was accomplished by TARRANT (*109*) using a platinum catalyst. The reaction was carried out in an autoclave at 225 to 250° for 16 hours and good yields were obtained. However, attempted additions of methyl dichlorosilane to $CF_2 = CFCl$ and perfluoroolefins with this technique resulted in violent explosions. MCBEE and PUERCKHAUER (*71*) succeeded in adding $HSiCl_3$ to $C_5F_{11}CF{=}CF_2$, using peroxide catalyst in an autoclave at a temperature of 125 to 130° for 60 hours. Other olefins which were added to $HSiCl_3$ include $C_2F_5CH{=}CH_2$, and $CF_3Cl = CCl_2$; when the latter olefin was used, the silane moiety added to the β-carbon atom and $CF_3\underset{\displaystyle SiCl_3}{\underset{|}{C}}{=}CCl_2$ was obtained.

c) The addition of fluoroalkyl halides to vinyl silanes was first reported in 1955 by HASZELDINE and MARKLOW (*48*) and TARRANT et al.

(*109*). HASZELDINE added CF_3I to vinyltrichlorosilane under thermal conditions and obtained $CF_3CH_2CHISiCl_3$. TARRANT added CF_2Br_2 and $CF_2BrCFClBr$ to $CH_2{=}CHSi(CH_3)_3$ using benzoyl peroxide and isolated $CF_2BrCH_2CHBrSi(CH_3)_3$ and $CF_2BrCFClCH_2CHBrSi(CH_3)_3$. Upon treatment with alkali, α,β-unsaturated derivatives were obtained. Further treatment of the butenyl silane $CF_2BrCFClCH{=}CHSi(CH_3)_3$ with zinc results in dimerization to the perfluorocyclobutane derivative,

$$\begin{array}{l}CF_2\text{—}CFCH{=}CHSi(CH_3)_3\\ |\quad\;\; |\\ CF_2\text{—}CFCH{=}CHSi(CH_3)_3\end{array}$$

as well as in the formation of $CF_2{=}CFCH{=}CHSi(CH_3)_3$. It is realized that while no silane has been produced by this method which can be construed to be a silicone monomer, it is conceivable that this reaction could be applied to make di- or trifunctional fluoroalkyl silanes. The presence of bromine or iodine in the alkyl group might be deleterious to the thermal stability of any resulting polymer.

d) Dimerization of allyl silanes with fluoroolefins results in the formation of cyclobutylmethylenesilanes in low yields (*44*). For example:

$$CH_2{=}CHCH_2Si(OC_2H_5)_2CH_3 + CF_2{=}CF_2 \xrightarrow[150^\circ/16\text{ hrs.}]{Bz_2O_2} \begin{array}{l}CH_2\text{—}CH\text{—}CH_2Si(OC_2H_5)_2CH_3\\ |\quad\;\;\; |\\ CF_2\text{—}CF_2\end{array}$$

e) The reaction of fluoroalkyl organometallic compounds with cloro-or alkoxysilanes have been this far confined to the use of Grignard and lithium derivatives. PIERCE, MCBEE and CLINE (*86*) reacted $C_3F_7CH_2CH_2MgBr$ with tetrachloro- and tetraalkoxysilanes and isolated $(C_3F_7CH_2CH_2)_2Si(OCH_2H_5)_2$. MCBEE and ROBERTS (*70*) obtained similar results with $C_3F_7CH_2CH_2MgCl$ and $Si(OR)_4$. HASZELDINE and MARKLOW (*48*) reported that the reaction of CF_3MgI and $SiCl_4$ resulted in the formation of $(CF_3)_2SiCl_2$ and CF_3SiCl_3. Simultaneous addition of C_3F_7I and Cl_2SiEt_2 to an etheral solution of methyllithium was reported by PIERCE, MCBEE, and JUDD (*87*) to give a low yield of $(C_3F_7)_2Si(C_2H_5)_2$.

Listed in Table 18 on some fluorine-containing silicone monomers which have been reported.

2. Polymerization Studies. Hydrolysis of fluorine-containing aryl di- and trifunctional silanes yield varying results. When 3-trifluoromethylphenyl trichlorosilane was added drop-wise to a mixture of water, toluene and *t*-amyl alcohol, a high viscosity sticky syrup resulted, which after aging at 116° for six days was converted to an ether-soluble glass (*84*). Similar treatment of bis(3-trifluoromethylphenyl) dichlorosilane resulted in a silicanediol, a cyclotrisiloxane and a cyclotetrasiloxane (*84*).

Table 18. *Fluorine-containing Silicone Monomers*

Monomer	b. p. [° C/mm.]	n_D (° C)	Ref.
$CHF_2CH_2SiCl_3$	104—5		111
$CF_3C(SiCl_3){=}CCl_2$	165—6		71
$CF_3CH_2CHBrSiF_3$	81	1.3434(23)	110
$C_2F_5CH_2CH_2SiCl_3$	127—7.8/75.8		71
$CF_3CH_2CH_2Si(CH_3)Cl_2$	76/85	1.3937(27)	14
$C_2F_5CH_2CH_2Si(CH_3)Cl_2$	52.5/40	1.3692(25)	14
$C_3F_7CH_2CH_2Si(CH_3)Cl_2$	146	1.3609(22)	14
$CF_3CH_2CH_2Si(CH_3)(OCH_3)_2$	96.5/251	1.3576(20)	70
$CF_3CH_2CH_2Si(OCH_3)_3$	79.8—80.5/87	1.3547(20)	70
$(C_3F_7CH_2CH_2)_2Si(OH)_2$	64.5 (m. p.)		86
$C_3F_7CH_2CH_2Si(CH_3)(OC_2H_5)_2$	75.9/23	1.3502—11(25)	70
$C_3F_7CH_2CH_2Si(OC_2H_5)_3$	92.5/25	1.211(25)	70
$(C_3F_7CH_2CH_2)_2Si(OC_2H_5)_2$	117.5/25		70
$(CF_2CF_2CH_2CHCH_2)_2Si(OC_2H_5)_2$ (ring: CF_2 joined to CH)	117/3		44
$4\text{-}FC_6H_4SiCH_3(OEt)_2$	87.5—8.0/4	1.4558(25)	43
$(FC_6H_4)_2SiCl_2$			106
$3\text{-}CF_3C_6H_4SiCl_3$	108/47	1.4678(25)	45
$3\text{-}CF_3C_6H_4Si(OC_2H_5)_3$	86.5—7.5/4		45
$4\text{-}CF_3C_6H_4SiCl_3$	118.7	1.3783(25)	45
$3\text{-}CF_3C_6H_4Si(CH_3)Cl_2$	115.5/50	1.4639(25)	61
$(3\text{-}CF_3C_6H_4)_2SiCl_2$	125/3	1.4884(25)	45
$(3\text{-}CF_3C_6H_4)_2Si(OC_2H_5)_2$	104—15/1		45
$[2,5\text{-}(CF_3)_2C_6H_3]_2SiCl_2$	110/3	1.4248(30)	59

Hydrolysis of various other trifluoromethylphenylchloro-or-alkoxylsilanes also yielded polysiloxanes with good thermal stability and resistance to strong oxidizing agents such as fuming nitric acid and sulfuric acid (*61, 84*). Hydrolysis of chlorosilanes containing two trifluoromethyl groups on the benzene ring yield only the silane diol (*52*). Low molecular weight polymeric oils can be obtained from these monomers upon treating the dichlorosilane with a mixture of $CuCl_2$ and PbO in acetonitrile (*52*). Silmethylene polymers have been reported to be made by means of a Wurtz-Fittig condensation of

$$\begin{array}{c} Cl \\ | \\ ClSiCH_2Cl \\ | \\ C_6H_4CF_3 \end{array} \quad (60).$$

Although Rochow (*97*) observed that Si—CF_2H and Si—CF_3 groups are cleaved by cold water, Haszeldine (*48*) reported that CF_3SiCl_3 forms a high melting resin when hydrolyzed in water.

Hydrolysis of $C_3F_7CH_2CH_2Si(OC_2H_5)_3$ for 24 hours yielded only a viscous oil which increased in viscosity when heated with concentrated sulfuric acid for 18 hours at 130—150°, however no resinous material was obtained (*68*). Hydrolysis of $(C_3F_7CH_2CH_2)_2Si(OC_2H_5)_2$ in aqueous ethanol for 21 hours at room temperature yielded the solid disiloxane. Hydrolysis of the same monomer in aqueous methanol resulted in the formation of the diol in good yields (*68, 86*). Further treatment of the diol with concentrated sulfuric acid for 20 hours at 140—165° gave a viscous oil (*68*). Hydrolysis of the same monomer with 6N HCl also gave an oily product (*85*).

Upon hydrolysis of monomers obtained from the addition of fluoro-olefins to allyl silanes, FROST (*44*) obtained resin products as well as liquid polymers with lubricating properties.

Copolymers of $(CH_3)_2SiCl_2$ with $(C_3F_7CH_2CH_2)_2Si(OC_2H_3)_2$ and with $C_3F_7CH_2CH_2Si(CH_3)(OC_2H_5)_2$ were prepared by CLARK (*17*) by co-hydrolysis in 10% HCl and subsequent polymerization with either sodium hydroxide or potassium hydroxide. An elastomeric gum resulted, which when compounded and cured, exhibited a swell of 69 to 140% in 70/30 isooctane-toluene as compared to a 250% swell of commercial silicone rubber in the same fluid.

Work by STUMP (*108*) led to the preparation of some cyclic siloxane derivatives of $CF_3CH_2CH_2Si(CH_3)Cl_2$, $C_2F_5CH_2CH_2Si(CH_3)Cl_2$, $C_3F_7CH_2CH_2Si(CH_3)Cl_2$, and $CH_3CH(CCF_3)CH_2Si(CH_3)Cl_2$. The tetramers

Table 19. *Cyclic Siloxanes*

Compound	b. p. [° C/mm.]	n_D (° C)
$[CF_3CH_2CH_2Si(CH_3)O]_4$	107—8/6	1.3663(20)
$[CF_3CF_2CH_2CH_2Si(CH_3)O]_3$	114—16/5	1.3554(20)
$[CH_3CH(CF_3)CH_2Si(CH_3)O]_3$	115—116/8	1.3830(20)
$[CF_3CF_2CF_2CH_2CH_2Si(CH_3)O]_4$	111—113/14	1.3447(20)

were obtained in the highest yield (42%), but other compounds were also isolated. Listed in Table 19 are some of the compounds obtained by aqueous hydrolysis.

Attempted homopolymerization of these cyclic siloxanes with either potassium hydroxide, sodium hydroxide or cesium hydroxide yielded only viscous oils. It is believed that the strong electronegative effect of fluorine accounts for the stability of the cyclic compounds to ring opening.

Table 20. *Properties of Silastic LS-53* (*43*)

Durometer (Shore A)	67
Tensile Strength (psi)	1180
Elongation (%)	270
Compression Set (% after 22 hours at 300° F)	30
% Swell after 24 hrs. in ASTM Fuel B	+20

A recent announcement by the Dow Corning Corporation of a fluoroalkyl silicone elastomer, called "Silastic LS-53" indicates that a suitable catalytic system has been found for polymerization of ring compounds of this type. This elastomer is comparable to ordinary silicone rubber in physical properties except that it exhibits a very low swell in fuels and oils.

IV. Homopolymers and Copolymers of Styrenes

1. Monomer Synthesis. The isomeric styrenes with one fluorine atom on the nucleus are all known. They can be prepared by reacting a fluorobenzaldehyde with methylmagnesium bromide followed by dehydration (*12A*). The reaction sequence is shown below:

$$F{-}C_6H_4{-}CHO + CH_3MgBr \rightarrow F{-}C_6H_4{-}CH(OH)CH_3 \xrightarrow[\text{hydroquinone}]{KHSO_4} F{-}C_6H_4{-}CH{=}CH_2$$

The preparation of *p*-fluorostyrene can be effected by reacting *p*-fluorophenylmagnesium bromide with acetaldehyde and dehydrating using either potassium hydrogen sulfate or phosphoric anhydride (*3*). MARVEL (*80*) reports that *m*-trifluoromethyl styrene can be synthesized by the following route.

$$CF_3{-}C_6H_4{-}MgBr \xrightarrow{CH_3CHO} CF_3{-}C_6H_4{-}CH(OH)CH_3 \xrightarrow{P_2O_5} CF_3{-}C_6H_4{-}CH{=}CH_2$$

This same procedure is used by RENOLL (*94*) for the preparation of *p*-fluorostyrene and *m*-trifluoromethylstyrene.

The synthesis of a number of fluorinated styrenes is described in a paper by BACHMAN and LEWIS (*3*). They react a fluoro- or trifluoromethylXphenylmagnesium bromide with the appropriate aldehyde or ketone as shown below, where X is F or CF_3.

$$X{-}C_6H_4{-}MgBr + R{-}\overset{O}{\overset{\|}{C}}{-}CH_2R' \longrightarrow X{-}C_6H_4{-}C(OH)(R){-}CH_2R' \longrightarrow X{-}C_6H_4{-}C(R){=}CHR'$$

Secondary and tertiary alcohols are dehydrated best with potassium bisulfate or phosphoric anhydride while potassium hydroxide is reported to be the best reagent for dehydration of primary alcohols. In a similar manner, *p*-fluoromethyl styrene can be prepared from *p*-fluorophenylmagnesium bromide and acetone (*103*).

Several styrenes can be prepared by decarboxylation of the corresponding cinnamic acids (*112*). This method is used for the preparation of *o*-fluorostyrene starting with *o*-fluorobenzaldehyde (*77*).

$$o\text{-}FC_6H_4CHO \longrightarrow o\text{-}FC_6H_4CH{=}CH{-}CO_2H \longrightarrow o\text{-}FC_6H_4CH{=}CH_2$$

A series of trifluoromethyl- and chloro-(trifluoromethyl)-substituted styrenes are reported by McBee and Sanford (*22*). These compounds are synthesized by reacting the appropriate Grignard reagent with acetaldehyde and the subsequent dehydration of the alcohol.

Dickey and Stanin (*39*) report the preparation of α-trifluoromethylstyrene and α-difluoromethylstyrene following the reactions below:

$$CH_3\overset{O}{\overset{\|}{C}}CF_3 + C_6H_5MgBr \longrightarrow C_6H_5\underset{CF_3}{\overset{OH}{C}}{-}CH_3 \longrightarrow C_6H_5\underset{CF_3}{C}{=}CH_2$$

The synthesis of α,β,β-trifluorostyrene and α-chloro-β,β-difluorostyrene is accomplished by the following reaction sequence (*19*) (where CX_3 is CF_3, $CClF_2$ or CHF_2):

$$CX_3\overset{O}{\overset{\|}{C}}Cl + C_6H_6 \xrightarrow[CS_2]{AlCl_3} C_6H_5\overset{O}{\overset{\|}{C}}CX_3 + HCl$$

$$C_6H_5\overset{O}{\overset{\|}{C}}CX_3 + PCl_5 \longrightarrow C_6H_5CCl_2CX_3 + POCl_3$$

$$C_6H_5CCl_2CX_3 + Br_2 + SbF_3 \longrightarrow C_6H_5CClFCX_3 \xrightarrow{Zn} C_6H_5CF{=}CF_2$$

$$C_6H_5CCl_2CX_3 \xrightarrow{Zn} C_6H_5CCl{=}CF_2$$

Alternate procedures for preparing α,β,β-trifluorostyrene are reported by Prober (*89*). Benzene and chlorotrifluoroethylene are reacted in a hot tube for two hours at 572—602° yielding the desired product. Either boron trifluoride, activated charcoal, tetramethyllead or ammonia are claimed to accelerate the reaction. This patent also describes the following reactions:

$$C_6H_5CH_2CHF_2 + PCl_5 \xrightarrow{CCl_3H} C_6H_5CCl_2CHF_2$$

$$C_6H_5CCl_2CHF_2 + SbF_3 \longrightarrow C_6H_5CClFCHF_2 \xrightarrow[dioxane]{Zn\ dust} C_6H_5CF{=}CHF$$

$$C_6H_5CClFCHF_2 \xrightarrow{KOH} C_6H_5CF{=}CF_2$$

In a recent paper, Dixon (*42*) reports the synthesis of several styrenes with fluorine substituents on the vinyl group by reacting phenyllithium with an excess of fluorine-containing olefin at $-80°$. For example:

$$CF_2{=}CF_2 + C_6H_5Li \longrightarrow C_6H_5CF{=}CF_2$$

$$CF_2{=}CCl_2 + C_6H_5Li \longrightarrow C_6H_5CF{=}CCl_2$$

The monomers described in this discussion are listed in Table 21 along with some physical properties.

Table 21. *Fluorine-containing Styrenes*

Substituent	b. p. [° C/mm.]	n_D (t ° C)	Ref.
o-Fluoro	32—4/3	1.5197(20)	12A
m-Fluoro	30—1/4	1.5173(20)	12A
p-Fluoro	29—30/4	1.5158(20)	12A,3
m-Trifluoromethyl	55/17	1.4655(20)	3
2,5-bis(Trifluoromethyl)	58/20	1.4237(25)	72
3,5-bis(Trifluoromethyl)	60/20	1.4220	72
α,β,β-Trifluoro	68—9/75	1.4741(20)	19, 42, 89
α-Chloro-β,β-difluoro	100/100	1.5080(20)	19
m-CF_3, α-CF_3	83—4/40	1.4625(25)	3
m-CF_3, β-CH_3	93—5/40	1.4724(25)	3
o-Br, *p*-CF_3	72—3/5	1.5228(20)	3
m-CF_3, *p*-F	77—8/40	1.4522(20)	3
m-CF_3, *p*-F, CH_3	89—91/40	1.4530(20)	3
p-F, CH_3	97.5—101.5/95	1.5120(20)	103
4-chloro-3-trifluoromethyl-	69/6	1.4980(25)	72
4-chloro-3-trifluoromethyl-methyl-	64/2.5	1.4944(25)	72
4-chloro-2-trifluoromethyl-	56/6	1.4919(25)	72
3,4-dichloro-5-trifluoromethyl-	75/3	1.5200(25)	72
3,4-bis(trifluoromethyl)-	78/20	1.4370(25)	72
4-chloro-2,5-bis(trifluoromethyl)-	78/20	1.4512(25)	72
4-chloro-3,5-bis(trifluoromethyl)-	68/6	1.4592(25)	72
α,β-difluoro	86.2—90.2/60	1.5061(20)	89
α-fluoro-β,β-dichloro-	101/12		42
α,β-difluoro-β-chloro-	174		42
β-trifluoromethyl-α,β-difluoro-	148		42

2. Polymerization Reactions. The most detailed polymerization study of nuclear substituted styrenes is that by MARVEL and coworkers (*78, 79*). Copolymerization of four fluorine-containing styrenes with butadiene is accomplished in a typical emulsion recipe at 50°. The styrenes entering the copolymer somewhat more easily than styrene itself. Evaluation of these copolymers in a tread stock recipe indicate that they are approximately equal to GR-S in physical properties and that *m*-fluorostyrene may possibly be superior. Some properties of these copolymers and the quality index of the vulcanizates are given in Table 22. Quality index is discussed by JUVE (*56*) and is defined as the observed flexures to the flexures of a similar GR-S tread stock having a hysteresis temperature rise equal to the observed temperature rise.

Homopolymers of *o*-fluoro- and *p*-fluorostyrene are hard, brittle, clear resins and can be prepared either using ultraviolet light (*112*) or benzoyl peroxide (*3*) catalysis. Monomers such as 1,5-difluoro-, 3,4-difluoro-, and 1-fluoro-5-chlorostyrene can be to homopolymerized and copolymerized with such monomers as methyl methacrylate maleic anhydride, isobutylene, butadiene, vinyl chloride, acrylonitrile, N-vinyl pyrrole, indene and N-vinyl carbazole (*13*). The copolymers are said to have excellent electrical properties.

The homopolymer of *m*-trifluoromethylstyrene is a resin melting at 195°, decomposing at 210° and forms clear colorless films with good heat resistance (*95*). When ultraviolet light is used as a catalyst, the homopolymer is benzene soluble and softens at 130 to 155° (*80*). RENOLL (*96*) reports that replacement of styrene with *m*-trifluoromethyl styrene in GR-S improves tensile strength and plasticity.

Table 22. *Fluorinated Styrene-Butadiene Emulsion Copolymers* (*56*)

Comonomer	Wt. Ratio Butadiene Styrene	Time [hrs.]	Conversion %	Benzene Solubility %	[η]	Fluorinated Styrene %	Copolym. Vulcanizate Quality Index	Control Vulcanizate Quality Index
o-Fluorostyrene . . .	75:25	10	83.5	100	1.90	27.2	0.7	1.6
							0.9	1.8
m-Fluorostyrene . .	75:25	12	70	96	1.44	24.3	4.2	1.8
p-Fluorostyrene . .	75:25	14	71	96	2.01	—	2.2	1.8
							2.7	2.0
m-Trifluoromethyl-styrene	75:25	14.5	75	100	1.83	26.8	2.4	2.4
							2.0	2.9
Styrene (control) . .	75:25	11	77	98	2.12	—	—	—

Resinous homopolymers can be prepared from both 4-fluoro-3-trifluoromethyl styrene and 2-bromo-4-trifluoromethyl styrene with benzoyl peroxide at 70°. Copolymers with butadiene can be prepared but no evaluation data are given (3). BACHMAN and LEWIS (3) also report that 2-fluoro-, 3-trifluoromethyl-, and 3-trifluoromethyl-4-fluoro-α-methylstyrene do not homopolymerize but will readily copolymerize with butadiene; β-methyl-3-trifluoromethylstyrene neither homopolymerizes or copolymerizes with butadiene.

The trifluoromethyl- and chlorotrifluoromethyl- substituted styrenes which are listed in Table 21 copolymerize with butadiene in emulsion at 50°; molecular weights range from 20000 to 56000. The homopolymerization and copolymerization reaction of *m*-trifluoromethyl-, 2,5-bis(trifluoromethyl)- and 3,5-bis(trifluoromethyl)styrenes with butadiene is discussed briefly by PIERCE and MCBEE (*84*). Determination of reactivity ratio of *m*-trifluoromethyl- and 2,5-bis(trifluoromethyl)styrene with styrene and methyl methacrylate is being studied in this Laboratory (*20*).

Homopolymerization of α,β,β-trifluorostyrene can be accomplished at 70—75° with benzoyl peroxide, yielding a polymer which softens at 181—187° (*89*).

LIVINGSTON (*65*) reports the homopolymerization of α,β,β-trifluorostyrene in a radical-catalyzed emulsion recipe or with anionic catalysts such as sodium, sodium in liquid ammonia and sodium methoxide. Films of the homopolymer are clear, tough, and flexible and X-ray studies indicate high orientation but no attendant crystallinity. The dielectric constant and loss tangent measurements are reported for both the powder and discs prepared by compression molding.

Copolymers can be obtained with styrene and trifluorochloroethylene. Reactivity ratios with styrene of $r_1 = 0.07$ and $r_2 = 0.66$ (where M_1 is trifluorostyrene) are obtained with a Q value of 0.37 and e value of +.95 calculated on the basis of the revised Q and e values of styrene, which are 1.0 and —0.8 respectively. The copolymerization curve of styrene-trifluorostyrene has a crossover at about 0.7 mole fraction of styrene and a strong tendency to alternate is also noted in this region.

V. Miscellaneous Polymers

Perfluoroacrylyl Fluoride. Perfluoroacrylyl fluoride does not homopolymerize with either peroxides, ultraviolet light or ionic catalysts. Emulsion polymerization is not possible because of hydrolysis of the monomer. Copolymers of perfluoroacrylyl fluoride with styrene and butadiene can be prepared but contain less than 10% of the fluorine-containing monomer. A tough plastic copolymer with vinyl acetate analyzed for 32 mole percent of the acrylyl fluoride. On heating, this polymer became rubbery at 75° and charred above 250°.

N-Perfluorovinyl piperidine (*9*). This monomer does not form homopolymers in bulk or emulsion and can not be copolymerized with perfluoropropylene.

Perfluoroazomethines (*6*). Perfluoro-2-aza-alkenes ($C_nF_{2n+1}N{=}CF_2$, where $n = 2, 3, 4$) hydrolyze readily in aqueous basic solutions and neutral solutions. Peroxides and ionic catalysts do not give a homopolymer.

Tetrafluoroallene (*54*). This monomer is prepared by the reactions shown below and boils at $-38 \pm .5°$.

$$CF_2{=}CH_2 + Br_2CF_2 \longrightarrow CF_2BrCH_2CF_2Br \longrightarrow CF_2{=}CHCF_2Br$$
$$CF_2{=}CHCF_2Br \longrightarrow CF_2{=}C{=}CF_2$$

At room temperature, under pressure, polymerization occurs in a few hours to yield first a liquid then a white solid.

Fluorinated Polyphenyls (*49*). Polyphenyls of the type

$$\left(-C_6F_4- \right)_n$$

where n is as high as 10, are prepared by reacting the corresponding diiodo or dibromo derivative at 200—250° with cooper as catalyst. The high molecular weight polymer (1300—1700) shows good thermal stability.

Bibliography

1. AHLBRECHT, A. H., and D. W. CODDING: Acrylates of difficultly esterified alcohols. J. Amer. chem. Soc. **75**, 984 (1953).
2. ANSPON, H. D., and J. J. BARON JR.: Development of a rigid transparent plastic material suitable for aircraft glazing at elevated temperature. Wright Air Development Center Technical Report 57-24 (1957).
3. BACHMAN, G. B., and L. L. LEWIS: Monomers and polymers. I. Fluorinated styrenes. J. Amer. chem. Soc. **69**, 2022—2025 (1947).
4. BARR, J. T., K. E. RAPP, R. L. PRUETT, C. T. BAHNER, J. D. GIBSON and R. H. LAFFERTY JR.: Reactions of polyfluoro olefins. III. Preparation of polyfluoro ethers. J. Amer. chem. Soc. **72**, 4480—4482 (1950).
5. BITTLES JR., J. A.: Fluorinated alkyl acrylates and methacrylates. U. S. Patent 2,628,958 (1953).
6. BORDERS, A. M.: Synthetic rubbers from carbon-fluorine compounds. Wright Air Development Center Technical Report 52-197, Part I (1951).
7. — Synthetic rubbers from carbon-fluorine compounds. Wright Air Development Center Technical Report 52-197, Part II (1952).
8. BOVEY, F. A.: Synthetic rubbers from carbon-fluorine compounds. Wright Air Development Center Technical Report 52-197, Part III (1953).
9. — Synthetic rubbers from carbon-fluorine compounds. Wright Air Development Center Technical Report 52-197, Part IV (1955).
10. — Synthetic rubbers from carbon-fluorine compounds. Wright Air Development Center Technical Report 52-197, Part V (1956).
11. —, J. F. ABERE, G. B. RATHMANN and C. L. SANDBERG: Fluorine-containing polymers. III. Polymers and copolymers of 1,1-dihydroperfluoroalkyl acrylates. J. Polymer Sci. **15**, 520—536 (1955).
12. — — Fluorine-containing polymers. IV. Polymers of acrylates of fluorine-containing ether alcohols. J. Polymer Sci. **15**, 537—543 (1955).

12A. BROOKS, L. A.: The preparation of substituted styrenes. J. Amer. chem. Soc. **66**, 1295—1297 (1944).

13. —, and M. NAZZEWSKI: Electrical insulation from dihalogenated polystyrene and copolymers. U. S. Patent 2,406,319 (1946).
14. BUTLER, G. B., R. DUNMIRE, G. W. DYKES and P. TARRANT: Preparation and polymerization of fluoroalkylmethylchlorosilanes. Abstracts 130th Meeting Amer. chem. Soc. 3 S (1956).
15. BUXTON, M. W., M. STACEY and J. C. TATLOW: Studies upon α-trifluoromethylacrylic acid, α-trifluoromethylpropionic acid and some derived compounds. J. chem. Soc. **1954**, 366—374.
16. CHANEY, D. W.: Perfluorinated acrylonitriles. U. S. Patent 2, 439, **505**, April 13, 1948.
17. CLARK JR., R. T.: Elastomeric fluoroalkyl silicone copolymers. Wright Air Development Center Technical Report 54-213 (1954).
18. CODDING, D. W., T. S. REID, A. H. AHLBRECHT, G. H. SMITH JR. and D. R. HUSTED: Fluorine-containing polymers. II. 1,1-dihydroperfluoroalkyl acrylates: preparation of monomers. J. Polymer Sci. **15**, 515—519 (1955).
19. COHEN, S. G., H. T. WOLOSINSKI and P. J. SCHEUER: α,β,β-trifluorostyrene and α-chloro-β,β-difluorostyrene. J. Amer. chem. Soc. **71**, 3439—3440 (1949).

20. Coleman jr., L. E.: Unpublished data.
21. Coleman, L. E., and G. R. Eykamp: Polymerization of some α,α,ω-trihydroperfluoroacrylates. Wright Air Development Center Technical Memorandum 57-56 (1957).
22. —, D. A. Rausch jr. and W. R. Griffin: Polymerization of some 1-alkyl-1-hydroperfluoroalkyl acrylates. Wright Air Development Technical Memorandum 57-57 (1957).
23. Coover, H. W., and J. B. Dickey: Polymerization of monomeric vinyl compounds. U. S. Patent 2,675,372 (1954).
24. Coover jr., H. W., J. E. Stanin and J. B. Dickey: Alkenyl trifluoroacetates. U. S. Patent 2,525,526 (1950).
25. Copenhaver, J. W.: Arctic rubber. The M. W. Kellogg Company. Report No. RL-54-353 (1954) and Report No. RL-55-401 (1955).
26. — Arctic rubber. The M. W. Kellogg Company. Report No. RL-55-434 (1955).
27. Corley, R. S., J. Lal and M. W. Kane: The properties of some fluorinated vinyl ethers. J. Amer. chem. Soc. **78**, 3489—3493 (1956).
28. Crawford, J. W. C., R. H. Stanley and *Imperial Chemical Industries Ltd.*: Fluoroalkyl methacrylate esters and polymers. British Patent 580,665 (1946).
29. — Fluorine-containing esters and their polymers. British Patent 616,849 (1949).
30. Dahlquist, C. A.: Synthetic rubbers from carbon-fluorine compounds. Wright Air Development Center Technical Report 52-197, Part VI (1955).
31. Darrall, R. A., F. Smith, M. Stacey and J. C. Tatlow: Organic fluorides. Part IX. The formation and resolution of α-hydroxy-α-trifluoromethylpropionic acid. J. chem. Soc. **1951**, 2329—2332.
32. Dickey, J. B.: α-(fluoromethyl)acrylamides. U. S. Patent 2,541,465 (1951).
33. — α-(fluoromethyl)acrylonitrile. U. S. Patent 2,541,466 (1951).
34. — Fluoroacrylate ester polymers. U. S. Patent 2,472,811 (1949).
35. — Derivatives of fluorinated methacrylic acid. U. S. Patent 2,472,812 (1949).
36. —, and H. W. Coover: Polymerization with alkyl phosphite catalysts. U. S. Patent 2,652,393.
37. —, and J. G. McNally: Copolymers of β,β-difluoroacrylates. U. S. Patent 2,571,687 (1951).
38. —, and T. E. Stanin: Polymers of α-fluoroacetoxy acrylonitrile compounds. U. S. Patent 2,464,120 (1949).
39. — — Polyfluoromethyl styrene. U. S. Patent 2,475,423 (1949).
40. — — Alkenyl fluoroacetates. U. S. Patent 2,525,530 (1950).
41. Dishart, K. T., and R. Levine: A new synthesis of ketones containing one perfluoroalkyl group. J. Amer. chem. Soc. **78**, 2268—2270 (1956).
42. Dixson, S.: Elimination reactions of fluoroolefins with organolithium compounds. J. org. Chem. **21**, 400—403 (1956).
43. *Dow Corning Corporation:* Data sheet, handling and processing Silastic LS-53. Midland, Michigan, 1957.
44. Frost, L. W.: Fluorine-containing organosilicon compounds. U. S. Patent 2,596,967 (1952).
45. — Trifluoromethylphenyl siloxanes. U. S. Patent 2, 636, 896 (1953).
46. Haas, H. C., E. S. Emerson and N. W. Schuler: Polyvinyl trifluoroacetate. J. Polymer Sci. **22**, 291—302 (1956).
47. Halpern, B. D., W. Karo, L. Laskin, P. Levine and J. Zomlefer: Synthesis of monomeric materials. Wright Air Development Center Technical Report 54-264 (1954).
48. Haszeldine, R. N., and R. J. Marklow: Fluoroalkyl compounds of silicon. Abstracts 128th Meeting Amer. chem. Soc. 43M (1955).

49. Hellmann, M., A. J. Bilbo and W. J. Pummer: Synthesis and properties of fluorinated polyphenyls. J. Amer. chem. Soc. **77**, 3650—3651 (1955).
50. Henne, A. L., and C. J. Fox: Ionization constants of fluorinated acids. III. Unsaturated acids. J. Amer. chem. Soc. **76**, 479—481 (1954).
51. Howk, B. W., and R. A. Jacobson: Vinyl fluoroacetate polymers. U. S. Patent 2,436,144 (1948).
52. Hyde, J. F.: Siloxanes. U. S. Patent 2,629,725 (1953).
53. Izard, E. F., and S. L. Kwolek: Reaction of perfluoroalkyl chlorides with silicon. J. Amer. chem. Soc. **73**, 1156—1158 (1951).
54. Jacobs, T. L., and R. S. Bauer: Tetrafluoroallene. J. Amer. chem. Soc. **78**, 4814—4816 (1956).
55. Jones, F. B.: Battelle memorial institute. Private Communication.
56. —, P. B. Stickney, L. E. Coleman jr., D. A. Rausch and A. M. Lovelace: Polymerization of some fluorine-containing olefin oxides. J. Polymer Sci. **26**, 81—88 (1957).
56A. Juve, A. E.: Crack growth in GR-S tread stocks. Ind. Eng. Chem. **39**, 1494—1498 (1947).
57. Knobloch, F. W.: Polymers and copolymers of N-1,1-dihydrofluoroalkyl acrylamides. J. Polymer Sci. **25**, 453—463 (1957).
58. —, and H. C. Hamlin: Polymers derived from dihydroperfluorobutyl acrylate. Wright Air Development Center Technical Report 55-25 (1955).
59. Kohl jr., C. F.: (Trifluoromethyl)phenylsilanols. U. S. Patent 2,640,063 (1953).
60. — Halohydrocarbon derivatives of silanes. U. S. Patent 2,530,202 (1950).
61. — Fluorocarbon substituted methylsiloxanes. U. S. Patent 2,571,090 (1951).
62. — Organosilicon compounds. U. S. Patent 2,640,066 (1953).
63. LaZerte, J. D., W. H. Pearlson, J. L. Rendall and T. J. Brice: Presented before the Fluorine Chemistry Symposium, 120th Meeting of the Amer. chem. Soc., New York 1951.
64. —, D. A. Rausch, R. J. Koshar, J. D. Park, W. H. Pearlson and J. R. Lacher: Perfluoroacrylonitrile and its derivatives. J. Amer. chem. Soc. **78**, 5639—5641 (1956).
65. Livingston, D. I., P. M. Kamath and R. S. Corley: Poly-α,β,β-trifluorostyrene. J. Polymer Sci. **20**, 485—490 (1956).
66. McBee, E. T., and T. M. Burton: The preparation and properties of 3,3,3-trifluoro-1,2-epoxypropane. J. Amer. chem. Soc. **74**, 3022—3023 (1952).
67. —, W. Marzluff and O. R. Pierce: The ionization constants of some fluorine-containing alcohols. J. Amer. chem. Soc. **74**, 444—446 (1952).
68. —, and O. R. Pierce: Fluorine-containing elastomers. Wright Air Development Center Technical Report 55-191, Part II (1953).
69. — — and D. D. Smith: The ionization constants of some new fluorine-containing acids. J. Amer. chem. Soc. **76**, 3722—3725 (1954).
70. —, C. W. Roberts, C. F. Judd and T. S. Chao: Reaction of Grignard reagents with alkoxysilanes at elevated temperatures. J. Amer. chem. Soc. **77**, 1292—1293 (1955).
71. — — and G. W. R. Puerckhauer: Addition of silicochloroform and methyl dichlorosilane to fluorine-containing olefins. J. Amer. chem. Soc. **79**, 2326—2329 (1957).
72. —, and R. A. Sanford: Preparation and polymerization of trifluoromethyl- and chloro-(trifluoromethyl)-substituted styrenes. J. Amer. chem. Soc. **72**, 4053—4055 (1950).
73. —, P. A. Wiseman and G. B. Bachmann: Perfluoro dibasic acids and derivatives. Ind. Engng. Chem. **39**, 415—417 (1947).

74. McGinty, L.: α-fluoroacrylic esters and polymers thereof. British Patent 590,015 (1947).
75. — α-fluoroacrylic esters and polymers. U. S. Patent 2,454,663 (1948).
76. Manowitz, B.: The industrial future of radiation chemistry. Nucleonics **11**, No. 10, 18—20 (1953).
77. Marvel, C. S., and D. W. Hein: *o*-methoxy-, *p*-benzyl-, *o*-fluoro-, and *o*-cyano-styrenes. Further examples of the disproportionation of phenylmethylcarbinols to ethylbenzenes. J. Amer. chem. Soc. **70**, 1895—1898 (1948).
78. —, G. E. Inskeep, R. Deanin, A. E. Juve, C. H. Schroeder and M. M. Goff: Copolymers of butadiene with halogenated styrenes. Ind. Engng. Chem. **39**, 1486—1490 (1947).
79. — — —, D. W. Hein, P. U. Smith, J. D. Young, A. E. Juve, C. H. Schroeder and M. M. Goff: Copolymers of butadiene with alkyl, aryl, alkoxyl, and phenoxyl styrenes. Ind. Engng. Chem. **40**, 2371—2373 (1948).
80. —, C. G. Overberger, R. E. Allen and J. H. Saunders: The preparation and polymerization of some substituted styrenes. J. Amer. chem. Soc. **68**, 736—738 (1946).
81. Miller, W. T.: Perhaloacrylonitriles from perhaloolefins. U. S. Patent 2,691,036. (1954).
82. Park, J. D., M. L. Sharrah and J. R. Lacher: Action of an alcohol on perfluorocyclobutene. J. Amer. chem. Soc. **71**, 2337—2339 (1949).
83. Passino, H. J., and L. C. Rubin: Organosilicon compounds. U. S. Patent 2,686,194 (1954).
84. Pierce, O. R., and E. T. McBee: Fluorine-containing elastomers. Wright Air Development Center Technical Report 52-191, Part I (1952).
85. — — Chlorine- and fluorine-containing compounds for nonflammable materials. Wright Air Development Center Technical Report 53-462 (1954).
86. — — and R. E. Cline: The synthesis of fluorine-containing organosilanes. J. Amer. chem. Soc. **75**, 5618—5620 (1953).
87. — — and A. F. Judd: Preparation and reactions of perfluoroalkyllithiums. J. Amer. chem. Soc. **76**, 474—478 (1954).
88. Postelnek, William: Air force polymer development program. Rubber World **543—545** (July 1957).
89. Prober, M.: Fluorinated styrenes. U. S. Patent 2,651,627 (1953).
90. Rathmann, G. B., and F. A. Bovey: Fluorine-containing polymers. V. Light scattering and viscosity study of poly-1,1-dihydroperfluorobutyl acrylate. J. Polymer Sci. **15**, 544—552 (1955).
91. Rausch, D. A., L. E. Coleman jr. and A. M. Lovelace: α,β-unsaturated ketones. I. The preparation and polymerization of perfluoroalkyl propenyl ketones. J. Amer. chem. Soc. **79**, 4983—4984 (1957).
92. —, A. M. Lovelace and L. E. Coleman jr.: The preparation and properties of some fluorine-containing epoxides. J. org. Chem. **21**, 1328—1330 (1956).
93. Reid, T. S., D. W. Codding and F. A. Bovey: Vinyl esters of perfluoro acids. J. Polymer Sci. **18**, 417—421 (1955).
94. Renoll, M. W.: Vinyl aromatic compounds. III. Fluorinated derivatives. J. Amer. chem. Soc. **68**, 1159—1161 (1946).
95. — Trifluoromethyl derivatives of vinyl aromatic compounds. U. S. Patent 2,313,330.
96. — Rubbery copolymers of (trifluoromethyl)vinyl aromatic compounds. U. S. Patent 2,469,845.
97. — An introduction to the chemistry of the silicones (2nd Ed.). New York: John Wiley and Sons, Inc. **1951**.

98. Sandberg, C. L., and F. A. Bovey: Fluorine-containing polymers. VI. Copolymerization of 1,1-dihydroperfluorobutyl acrylate. J. Polymer Sci. **15**, 553—557 (1955).
99. Schweiker, G. C., and P. Robitschek: Condensation polymers containing fluorine. I. Synthesis of linear polyesters from fluorine-containing diols. J. Polymer Sci. **24**, 33—41 (1957).
100. —, R. R. White and R. N. Deleo: Investigation of condensation type elastomers. Wright Air Development Center Technical Report 55-221, Part I (1955).
101. — — — Investigation and condensation type elastomers. Wright Air Development Center Technical Report 55-221, Part II (1956).
102. —, B. S. Marks, R. R. White and R. N. Deleo: Investigation of condensation type elastomers. Wright Air Development Center Technical Report 55-221, Part III (1956).
103. Seymour, D., and K. B. Wolfstirn: Substituted styrenes. III. The preparation of some *m*- and *p*-substituted α-methylstyrenes. J. Amer. chem. Soc. **70**, 1177—1179 (1948).
104. Simons, J. H., and R. D. Dunlap: Silicon halides containing fluorocarbon radicals. U. S. Patent 2,651,651 (1953).
105. Smith, D. D., R. M. Murch and O. R. Pierce: Fluorine-containing polyethers. Ind. Engng. Chem. **49**, 1241—1246 (1957).
106. Sommer, L. H., E. W. Pietrusza and F. C. Whitmore: Peroxide-catalyzed addition of trichlorosilane to 1-octene. J. Amer. chem. Soc. **69**, 188 (1947).
107. Stedry, P. J., J. F. Abere and A. M. Borders: Fluorine-containing polymers. VII. 1,1-dihydroperfluoroalkyl acrylates: compounding and properties of the vulcanizates. J. Polymer Sci. **15**, 558—574 (1955).
108. Stump jr., E. C.: Preparation and polymerization of some fluoroalkylcyclosiloxanes. Wright Air Development Center Technical Report 56-493 (1957).
109. Tarrant, P., G. W. Dykes, F. F. Norris and D. E. O'Conner: The addition of perhaloalkanes to vinyl silanes. Abstracts 128th Meeting American Chemical Society 47M (1955).
110. Wagner, G. H.: Reactions of silanes with aliphatic unsaturated compounds. U. S. Patent 2,637,738 (1953).
111. Walborsky, H. M., and M. Schwarz: Addition Reactions to ethyl γ,γ,γ-trifluorocrotonate. J. Amer. chem. Soc. **75**, 3241—3243 (1953).
112. Walling, C., and K. B. Wolfstirn: Substituted styrenes. I. The decarboxylation of substituted cinnamic acids. J. Amer. chem. Soc. **69**, 852—854 (1947).
113. *Westinghouse Electric International Co.:* British Patent 706,703 (1954).

Fortschr. Hochpolym.-Forsch., Bd. 1. S. 114—158 (1958)

The Dynamic Mechanical Properties of High Polymers at Low Temperatures

By

A. E. Woodward and J. A. Sauer
College of Chemistry and Physics, the Pennsylvania State University

With 18 Figures

Table of Contents

I. Introduction

It is the purpose of the present article to review and discuss investigations of deformational behavior of high polymers in the temperature range below the primary transition. Our concern is thus with the mechanical properties of polymers in their "glassy" or crystalline state and not with polymers in the rubbery state or in solution[1]. Even with this restriction, the available literature is extensive, as within the last five years or so there has been a surprisingly large number of experimental

[1] There have recently become available excellent and extensive review articles by Ferry (1956) on the mechanical properties of polymeric materials in solution as well as over the transition region from the "glassy" to the rubbery state and by Treloar on the mechanical properties of rubbery materials.

investigations concerning the temperature dependence of the dynamic mechanical behavior of rigid or solid polymers below their glass-like transition temperature.

For most polymers or polymer systems, whether amorphous or crystalline, it is not sufficient to describe their mechanical behavior purely in terms of a glassy state below some transition zone in which segmental mobility is "frozen-in", and a rubbery or liquid state above this transition zone where segmental mobility is high and essentially unhindered. Rather, the data indicate that in many polymers multiple dispersion regions exist when we explore the mechanical response over a broad enough range of temperature or of frequency. These transition regions, during which the modulus or compliance changes relatively rapidly with temperature and the mechanical loss exhibits maxima, occur even in the so-called glassy state and, depending on the frequency, may be observed at temperatures in the vicinity of 150° K. Thus, even at low temperatures, the deformational response of many polymers is viscoelastic in nature and the polymer cannot be treated either as a rubber or as a simple elastic solid.

Most of the earlier investigations which have been made on rigid polymers have been carried out over a limited temperature range or else the results are given for room temperature only or for a few other selected temperatures nearby. In many of the recent investigations, however, the temperature range explored extends from 80° K to the melting point of the polymer and the temperature is continuously varied in between. Thus the body of accumulated knowledge concerning the mechanical and viscoelastic behavior of polymers from liquid nitrogen temperatures to their softening temperatures has been steadily increasing. It is desirable to review the existing literature on the subject both to assess the progress which has been made in our knowledge and understanding of the dynamic mechanical properties of polymers in their rigid state, and to delineate some of the areas in which additional study is badly needed. It is hoped also that a review of the experimental data will be a valuable aid to the development of theories of the viscoelastic properties of polymers in the region below their softening temperatures and will be helpful in correlating mechanical with other types of behavior such as dielectric, thermal, magnetic, and the like.

This review will be primarily concerned with the time dependent properties and in particular with the results of dynamic mechanical measurements of both amorphous and partially crystalline high polymers. Because of the comparatively high rigidity ($\sim 10^{10}$ dynes/cm^2) of most polymers below their primary transition temperatures, very few investigations have been made at low temperatures by such means as creep or stress-relaxation measurements.

It is appropriate, before considering the results of the specific experimental investigations, to give a short discussion of the general structure of high polymers and of the deformational response to be expected of simple viscoelastic materials. This is done in the following section which also includes a brief description of some of the experimental methods utilized for investigating the dynamic mechanical properties. The experimental data for amorphous polymers are then presented and discussed in Section III and for partially crystalline polymers in Section IV.

II. Mechanical Behavior of Simple Viscoelastic Substances

It has been shown by many investigators that in the range of small strains a high polymer may be represented in a first approximation by a mechanical model consisting of an instantaneous elastic element, a series of retarded elastic elements, and a flow element. The instantaneous elastic element is analogous in the real polymer to the elastic deformation associated with bond stretching and bond angle distortion, the retarded elastic elements are representative of the delayed elastic response associated with configurational elasticity of various segments of the polymer chains, and the flow element is representative of chain slipping, resulting in permanent deformation. Since in this article, our concern is only with the deformational changes occurring in polymers in their rigid or "glassy" state, the flow element may be deleted from the approximate model without loss of generality.

The model mentioned above has been developed to qualitatively describe the mechanical behavior of an amorphous polymer in which the long flexible polymer chains, consisting of repetitive monomer units, are not aligned in long range order but in which the individual chain units of the polymer molecule, under the action of thermal agitation, wander about from position to position. Clearly it cannot adequately represent the behavior of a crystalline polymer, such as polyethylene, in which the X-ray diffraction picture gives not only broad and diffuse halos characteristic of amorphous materials but also shows relatively sharp rings and intensity maxima characteristic of precise spacings of the atoms. Many other high polymers also show a pronounced tendency to "crystallize". In general the crystallites are very small in size and a given polymer molecule may wander through a number of alternate regions of order and disorder.

In such partially crystalline polymers, the viscoelastic behavior would be expected to greatly depend on the relative amounts of amorphous and crystalline regions and on their particular geometrical arrangement. Further, since the demarcation line between ordered and disordered regions is not sharp, there are regions of varying degrees of order throughout the polymer and hence the total mechanical behavior could hardly be represented by a single model which did not take this change of structure into account. Also in crystalline polymers, phase changes may occur which are not completely reversible and hence the model itself would have to change accordingly if it were to properly represent the real conditions.

Despite these limitations on the use of any model to represent the mechanical behavior of a high polymer, it is helpful and instructive, before examining the experimental results on the variation of the dynamic mechanical properties with frequency or temperature, to see what general behavior we might expect from the approximate model when it is subjected to a sinusoidal applied stress. For this purpose, we consider the simplest type of model in which there is only one retarded element and therefore only one relaxation time associated with internal molecular rearrangements. The results can later be generalized, if desired, to include any number of retarded elastic elements corresponding to repositioning under stress of each different size of molecular segment.

1. Mechanical Behavior of a Three Parameter Model

With the restrictions noted above, the model in question can now be represented by either the elements shown in Fig. 1A or Fig. 1B. Mathematically these two models are completely equivalent provided the moduli G_1 and G_2 and the viscosity η_1 of model B are properly related to the moduli G_0 and G and the viscosity η of model A. In model A, G_0 represents the instantaneous elastic response and for most polymers has a value of 10^{10} to 10^{11} dynes/cm². G, the retarded modulus, and η, the viscosity of the dashpot, are related to τ, the relaxation time, by the expression $\tau = \eta/G$. When a constant stress, σ_0, is applied to such a model the total strain ε will be made up of the sum of two parts

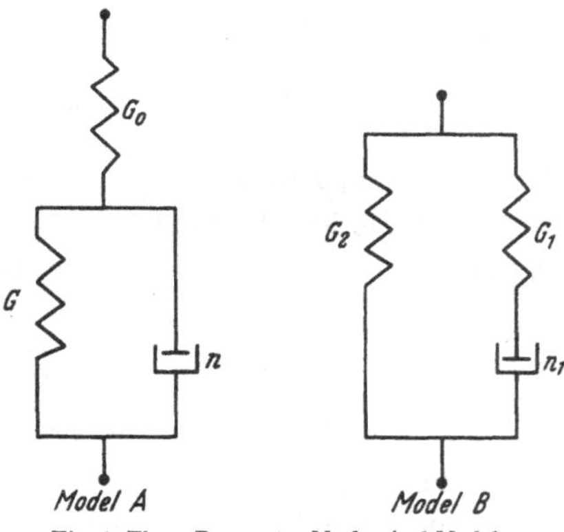

Fig. 1. Three Parameter Mechanical Models

$$\begin{aligned} \varepsilon &= \varepsilon_{el} + \varepsilon_{an} \\ &= \frac{\sigma}{G_0} + \frac{\sigma}{G}\left(1 - e^{-t/\tau}\right) \end{aligned} \tag{1}$$

where ε_{el} is the instantaneous elastic strain and ε_{an} the anelastic or time dependent strain. If the viscosity is high (corresponding to a low temperature) the retardation time τ is very long and hence over a reasonable experimental time period the response is almost elastic. If the viscosity is small (high temperatures), τ is small and the anelastic strain reaches its equilibrium value.

If a sinusoidal stress of frequency $\omega\,(\sigma = \sigma_0\, e^{i\omega t})$ is applied to this model the resulting deformational behavior may be discussed in terms of the storage and loss compliance J' and J'', where the complex compliance $J^* = J' - i\,J''$, or in terms of the storage and loss moduli G' and G'', where the complex modulus is $G^* = G' + i\,G''$. The storage, or in-phase component of the compliance is given by

$$J'(\omega) = \frac{1}{G_0} + \frac{1}{G}\,\frac{1}{1 + \omega^2\tau^2} \tag{2}$$

and the loss, or out-of-phase component of the compliance by

$$J''(\omega) = \frac{1}{G}\,\frac{\omega\,\tau}{1 + \omega^2\tau^2}\,. \tag{3}$$

The appropriate expression for the real and imaginary parts of the complex modulus can be obtained directly from these equations by noting that

$$G^* = \frac{1}{J^*}\,, \qquad G' = \frac{J'}{J'^2 + J''^2} \qquad \text{and} \qquad G'' = \frac{J''}{J'^2 + J''^2}\,.$$

The energy loss (ΔW) per cycle is directly related to either the loss compliance or the loss modulus by the following expression

$$\Delta W = \pi\,\sigma_0^2\,J''\,. \tag{4}$$

It is customary from many of the experimental arrangements to obtain values of the loss tangent ($\tan\delta$) or the so-called inverse $Q\,(Q^{-1})$ of the system. For small values of the damping these quantities are given by

$$Q^{-1} = \tan\delta = \frac{G''}{G'} = \frac{J''}{J'} = \frac{G_0}{G + G_0}\,\frac{\omega\,\tau}{1 + \dfrac{G}{G + G_0}\,\omega^2\tau^2}\,. \tag{5}$$

From these equations one perceives that the energy loss per cycle or the loss tangent will be very small at low values of $\omega\,\tau$, will rise to a maximum at some intermediate value, and will then fall again to a very low values at large values of $\omega\,\tau$.

The real part of the complex modulus or complex compliance does not go through a maximum. As can be seen from Eq. (2), the compliance J' will be large at low values of $\omega\,\tau\,(\sim 1/G)$, will decrease sharply at intermediate values, and will then tend to approach the constant value of $\left(\frac{1}{G_0}\right)$ as $\omega\,\tau$ becomes large. The real part of the dynamic modulus G' will show the opposite behavior with change of $\omega\,\tau$.

2. Mechanical Behavior of Real Polymer Systems

The model discussed above predicts that there should be one transition region where, as the sinusoidal stress frequency, ω, is increased, the loss tangent or internal friction should pass through a maximum and the storage compliance or modulus should undergo a relatively more rapid change with frequency. This is qualitatively the behavior observed experimentally in real polymers, both for amorphous and partially crystalline types, except that in real polymers there may well occur more than one such transition region and the specific nature of the changes in the vicinity of the transition region may not agree with that predicted by the simple model. One of the reasons for this discrepancy is that the three parameter model allows for only one relaxation mechanism. In an actual polymer with a wide range of molecular weights and with a wide range of internal structure in both the microscopic and macroscopic sense, one would expect that a wide range of relaxation times would be present and hence a more sophisticated model needed to more accurately represent its deformational properties.

Much progress has been made along these lines and considerable insight into the deformational behavior, particularly of true, amorphous polymers, has been obtained by use of models having an infinite set of retarded elastic elements in series. One can in fact frequently infer from the observed variation of the real and imaginary parts of J^* or G^* with frequency the shape of the relaxation or retardation spectrum of the particular polymer and from the distribution functions predict other types of mechanical behavior such as creep or stress relaxation. Ferry (1956) has written an extensive review on this subject. However almost all of the work carried out on calculations of the retardation or relaxation functions refer to the primary or glass-like transition and very little attention has been given to calculating such functions for secondary transitions.

3. Effect of Temperature on Mechanical Behavior

Both the simple model A with one relaxation time and a more sophisticated model with an infinite number of relaxation times provide for retarded or configurational elasticity and predict the occurrence of dispersion in the modulus and maxima in the internal friction if observations are made as a function of frequency at constant temperature. Furthermore, since τ is a measure of the time required to approach equilibrium and in the actual polymer is thus inversely proportional to the rate of segmental diffusion, it is expected that τ will decrease with increasing temperature. For some types of relaxation processes, it is customary to assume that, in accordance with the Arrhenius equation, τ varies exponentially with temperature:

$$\tau = \tau_0 \, e^{\Delta H/RT} \qquad (6)$$

where ΔH is the activation energy for the rate process involved and τ_0 may itself vary slowly with temperature. The decrease in τ — or increase in segmental diffusion rate — with temperature is a result both of thermal expansion leading to greater free volume and also of greater probability of segmental jumps over energy barriers because of the higher thermal energies. Hence it is to be expected that if frequency is kept constant, or nearly so, and temperature varied, dispersion phenomena will again arise.

This indeed is what is found experimentally when one measures the dynamic mechanical properties of polymers over a broad temperature range. One finds certain "transition" temperature regions where the modulus, as the temperature is raised, undergoes relatively sharper reductions. Experimentally, it is also noted that the damping rises to a maximum value and then falls again to a lower value. The chief differences between the observed behavior and the behavior predicted by the simple model A are that the shape of the observed dispersion indicates a range of relaxation times rather than a single time, and secondly, that frequently more than one relaxation mechanism is present with relaxation times of widely different orders of magnitude.

For completely amorphous polymers it has been found possible in a good number of instances to correlate the observed mechanical data by means of a "reduced variables" type of treatment as proposed by FERRY (1950). The assumption on which this method is based is that all relaxation times present vary in the same manner with temperature. If this assumption is met then all the observed data regardless of the actual temperature or the frequency of measurement can be plotted on a single curve and from this curve one can obtain the relaxation distribution function or the retardation distribution function.

4. Experimental Methods of Measuring Dynamic Mechanical Properties

To extend the time scale of damping and dynamic modulus measurements, one usually resorts to the use of a number of different methods. Generally, dynamic mechanical measurements for which both stress and the resulting strain are sinusoidal functions of time can be divided into four categories, that in some cases overlap, covering in all a frequency scale of about 0.1 to 10^8 cps. These four categories, discussed briefly in the following are: a) free vibration methods, b) forced vibrations in resonance, c) non-resonance forced vibration methods, and d) wave-propagation methods. These four general types give different measures of damping which, in general, can only be equated at low damping values.

Methods involving free vibrations usually include an added mass and cover the frequency range from 0.1 to 10 cps. The most commonly used apparatus in this group is the torsional pendulum giving the experimental quantities in terms of shear. The damping is obtained by measurement of the decay in amplitude, the logarithmic decrement, Δ, being determined. At low values of the damping, Δ is related to other measures of damping by the following relationships:

$$(\tan\delta)_s = G''/G' \cong \Delta/\pi \tag{7}$$

where $(\tan\delta)_s$ is the loss angle, measured in shear.

Forced vibration measurements at resonance, made either with or without added mass, are carried out at the resonant frequency of a free or loaded sample. If no additional mass is present the total mass of the system is that of the sample. In this case the resonant frequency is dependent on the sample dimensions and density. The resonance vibration of non-loaded specimens has been carried out by the transverse vibration of reeds, thin strips and rods, and the bending, longitudinal or torsional vibrations of rods. This general type of measurement covers a frequency range from approximately 10 to 10^5 cps. The damping or internal friction, Q^{-1}, is found from the half-width of the resonance curve and is given by $Q^{-1} = \frac{\Delta f}{f_0}$ where Δf is the half-width and f_0 is the resonance frequency. Only for small values of the damping can Q^{-1} be related to the log decrement by the simple relation: $Q^{-1} \cong \frac{\Delta}{\pi}$. The modulus is calculated from the resonant frequency f_0.

Non-resonance forced vibration methods can be carried out either with or without added mass. The frequency range attainable by these measurements is about 10^{-2} to 10^4 cps. The damping is obtained from the phase angle. Furthermore, if a non-resonance method is employed a larger frequency range can be measured than for the resonance cases; it is also simpler experimentally to keep the frequency constant while varying the temperature.

Wave propagation methods are especially applicable at high frequencies ranging from 10^5 cps to about 10^8 cps. The measure of damping for this type is the attentuation, α, given in units of cm^{-1}. For small damping and the propagation of shear waves $G''/G' \cong \frac{\alpha\lambda}{\pi}$ where λ is the wave length of the propagating wave. The storage modulus is obtained from the wave velocity.

III. Amorphous Polymers

In this section we will be principally concerned with the dynamic mechanical properties of non-crystalline high polymers. As is well known, these materials can be either in a glassy, rubbery or liquid state depending on such factors as temperature, molecular weight and amount of crosslinking. For amorphous polymers the principal change in the dynamic modulus or loss factor occurs over the temperature region within which the polymer is transformed from a glassy solid to a rubbery material or a liquid. This is usually perceived quite readily by the experimental methods employed for these investigations.

To date, secondary dispersion regions exhibiting corresponding smaller decreases of dynamic modulus with increasing temperature have been observed with a variety of amorphous polymers. However, except in a few cases, principally poly-(methyl methacrylate), the majority of systematic studies have been carried out to determine the effect of various agents, such as plasticizers, on the primary transition. Also, of the individual polymer samples studied generally little has ever been reported with regard to their molecular weight, branching content, chemical analysis, etc. which automatically precludes any quantitative comparison of data derived from different laboratories. Therefore any correlations attempted below must be considered semi-quantitative at best. Although it may be expected that thermal history, presence of water or other low molecular weight compounds, molecular weight, degree of branching or cross-linking, etc. may effect the secondary dispersion, with only a few exceptions, this information has not been given up to the present.

1. Polystyrene

Polystyrene has been available until recently only as a completely amorphous polymer. Hence all observations reported herein concerning the mechanical properties of this material apply only to the amorphous variety and not to crystalline polystyrene.

The most complete measurements on the mechanical behavior of polystyrene below its second order transition temperature would appear to be the low frequency measurements (~ 1 cps) of SCHMIEDER and WOLF (1953) and JENCKEL, the audio frequency measurements (~ 1000 cps) of SAUER and KLINE and the ultrasonic measurements of YAMAMOTO and WADA (10^5 cps) and of THURN (2×10^6 cps). All of these investigations included some low temperature measurements and in several instances, modulus and internal friction curves are given for the temperature range from 80° K to the softening region.

The most striking feature of the test results is the occurrence of a primary dispersion region, wherein the material passes from the hard glassy state ($G \sim 10^{10}$ dynes/cm^2) to a soft rubbery state ($G \sim 10^7$ dynes/cm^2) and the damping or internal friction exhibits a high maximum. The exact location of this transition, usually given as 370° K or a little higher, depends on the purity of the sample and the presence of low molecular weight ingredients. The higher transition temperature value for polystyrene compared to that of true rubbers, having comparable transitions at about 220° K, is considered to be a result of the increased chain stiffness arising from the presence of the bulky and heavy phenyl group attached to each monomer unit.

The effect of molecular weight on the mechanical properties in the transition region has been studied by MERZ, NIELSEN and BUCHDAHL. They showed that the shape of the primary dispersion region was not affected by a change in molecular weight from 100000 to 1000000.

The position of the modulus dispersion and of the internal friction peak is however definitely shifted by the presence of low molecular weight ingredients or by adding varying concentrations of a comonomer. For example, it has been shown by SCHMIEDER and WOLF (1953) that 1% of dimer, trimer, etc., will lower the α or primary transition of polystyrene by 16°. These same authors also show that on addition of varying concentrations of isobutylene to styrene to form a copolymer, the temperature of the damping maximum, which they found to occur at 390° K (0.9 cps) for the pure polystyrene, decreased linearly with isobutylene concentration until it reached the value 225° K for the pure polyisobutylene.

Secondary dispersion regions occurring at lower temperatures are found when mixtures of polystyrene and other compounds are studied. For example, BUCHDAHL and NIELSEN find that a mixture of polystyrene and a butadiene-styrene copolymer exhibits two peaks in mechanical loss, one peak arising from the polystyrene phase and the other from the rubber phase. These same authors also studied a graft copolymer of these two materials and found that the effect of the styrene-graft was to shift the rubbery transition temperature from 200° K to 250° K. Graft copolymers of polystyrene and rubber have also been studied by BLANCHETTE and NIELSEN. They find, for a sample containing 10% of GRS rubber, a damping peak at 230° K in addition to the 373° K peak characteristic of the polystyrene. A small secondary maximum has also been found by BUCHDAHL and NIELSEN in a sample of polystyrene of high molecular weight to which a low molecular weight fraction ($\sim$ 2000) was added.

The experimental results thus indicate that when polystyrene is copolymerized by a uniform chemical method, the mechanical loss shows but a single main peak located between the temperatures at which this transition would occur for the separate ingredients. However, if the specimen is formed from a physical mixture of the two ingredients or if it is a graft type copolymer, then the mechanical behavior is characterized by two transitions, one occurring at a temperature somewhat above the softening temperature of the second constituent, and the other occurring at the usual glass transition temperature for polystyrene. The graft copolymer thus behaves mechanically in a similar fashion to a physical mixture of two distinct phases despite the fact that, in this case, the phases cannot be separated.

Whether secondary dispersion regions do occur also in the pure polymer is uncertain. Both the data of SCHMIEDER and WOLF (1953) and that of SAUER and KLINE show that there are no "major" transitions encountered at any temperature from 80° K to the primary transition. However there are indications in the data of SCHMIEDER and WOLF (1953) that two small secondary transitions, showing only slightly increased compliance with increasing temperature, might occur. These might be referred to as a β transition occurring at about 320° K (8 cps) and a γ transition at about 130° K (11.3 cps). There is no β transition evident in the audio-frequency data of SAUER and KLINE, in the ultrasonic velocity data of SUBRAHMANGAN, in the high frequency measurements (1 to 6 megacycles/sec) of KRISHNAMURTHI and SASTRY, or in the torsional pendulum measurements of JENCKEL. On the other hand there is some slight indication in the high frequency data (2×10^6 cps) of THURN of a secondary transition in this temperature range and the mechanical loss values of YAMAMOTO and WADA, using a composite oscillator technique, do give a clearly distinguishable maximum (Fig. 2).

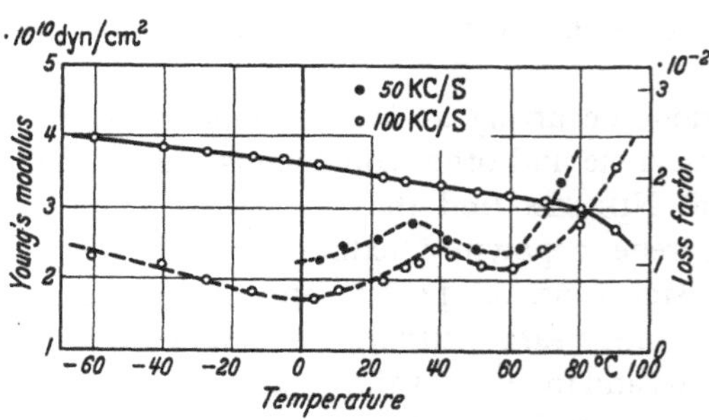

Fig. 2. Young's modulus (solid line) and loss factor (broken line) for polystyrene. Data of YAMAMOTO and WADA

With regard to the possible existence of a still lower temperature γ peak, the only other evidence that is available is that both the audio frequency measurements of SAUER and KLINE and the ultrasonic measurements of THURN show an apparent minimum in the loss curves at about 200 and 220° K respectively with progressingly slightly higher values of loss as the temperature is further decreased.

No molecular interpretation has yet been advanced for either the existence of a β or a γ dispersion region in polystyrene. The structure of polystyrene is such that there is no easily movable side group and diffusion of main chain segments under stress from one configuration to another would be expected to be very difficult, except at high temperatures, because of the heavy and bulky benzene ring. It would appear that further experimental study on highly purified and carefully prepared specimens is in order.

SAUER and KLINE (1956) studied poly-(alpha methyl styrene) in the audio frequency range and found no additional transitions from 80° K other than the one due to primary softening occurring above 370° K.

2. Polyisobutylene, Natural Rubber and Related Polymers

Polyisobutylene has been the subject of exhaustive investigation as reported by MARVIN in the region of and above the primary (α) transition at about —50° C. Investigations have also been made down to about 100° K in the frequency ranges of around 10 cps by SCHMIEDER and WOLF (1953), 30 to 300 cps by THOMAS and ROBINSON and 0.5×10^6 to 3.5×10^6 cps by KABIN and MIKHAILOV. With the possible exception of a small hump on the lower temperature side of the main softening dispersion apparent in the data of THOMAS and ROBINSON, no secondary dispersion regions have been reported by these three sets of investigators. Unpublished work of DEELEY in the 1000 cps region confirms this absence of secondary dispersions below a temperature of 220° K. FITZGERALD, GRANDINE and FERRY, however, have reported that the main dispersion region in polyisobutylene is actually a double peak. The appearance of this double peak has not been clearly explained but has been tentatively associated by FERRY, GRANDINE and FITZGERALD with the steric hindrance known to be present in the polyisobutylene molecule.

SCHMIEDER and WOLF have studied polybutadiene containing 70% 1,2 units and 30% 1,4 units, natural rubber, and butyl rubber in the temperature range from about 100° K up. These authors found no secondary transition for butyl rubber, as is also the case with its principal constituent polyisobutylene, but did report one at 173° K and 5.8 cps for polybutadiene and two for unvulcanized natural rubber at 188° K and 138° K at a frequency of about 10 cps. Upon sulfur vulcanization the dispersion regions in natural rubber seem to disappear at low percentages of sulfur, a peak reappearing around 170° K at higher sulfur amounts. No explanation for these secondary peaks has been given. Their confirmation awaits further investigations.

SCHMIEDER and WOLF (1953) have studied polyisobutylene of molecular weights of about 5×10^5, 9×10^5 and 1.75×10^6 and find only slight changes in the loss spectra.

3. Poly-(Methyl Methacrylate) and other Poly-(Acrylate Esters)

By far the largest number of dynamic mechanical property studies on the polyacrylates have been carried out on poly-(methyl methacrylate). This polymer has been investigated at frequencies as low as 10^{-3} cps to those as high as 6×10^6 cps over a wide temperature range. In addition to the main softening process in poly-(methyl methacrylate) at around 370° K or greater, one secondary maximum and possibly two exist at lower temperatures. The first secondary maximum reported, designated as the β peak, is readily apparent from the data of HEIJBOER in Fig. 3 where tan δ is given as a function of temperature at frequencies of .4, 2, 20, 200 and 2000 cps. As the frequency is increased both the α and β damping maxima shift to higher temperature; the β peak, however, shows a greater shift, merging with the α peak at higher frequencies. As a further illustration of this frequency dependence, a "transition" map of peak temperature *vs* log frequency including a majority of the existing data for the secondary maximum found for this polymer is given in Fig. 4. It appears from this figure that the maxima reported at frequencies of $.5 \times 10^5$, 10^5 and 2×10^5 cps by YAMAMOTO and WADA are not related to the β process defined by data at lower frequencies and that two damping maxima do occur. However, neither HOFF, ROBINSON and WILLBOURN, working in the 100—500 cps frequency range, nor SCHMIEDER and WOLF (1953), employing frequencies of .1—10 cps at temperatures as low as 120° K have found this dispersion. MIKHAILOV, on the other hand, has reported a dielectric

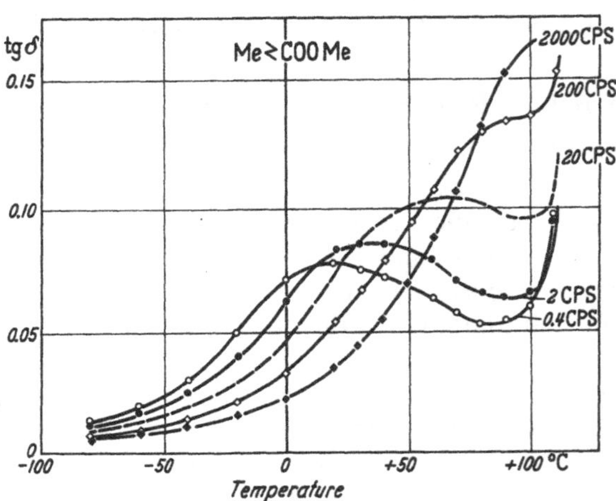

Fig. 3. Mechanical damping of poly-(methyl methacrylate) as a function of frequency for different temperatures. Data of HEIJBOER

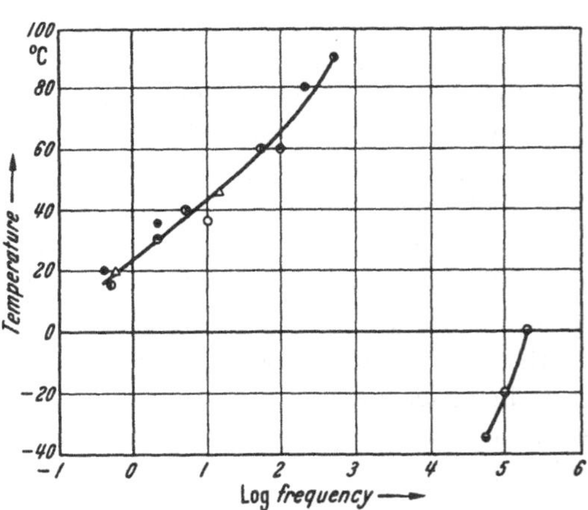

Fig. 4. Transition map for mechanical loss measurements of poly-(methyl methacrylate). Data of HEIJBOER (●), SCHMIEDER and WOLF (○), DEUTSCH, HOFF and REDDISH (◒), IWAYANAGI and HIDESHIMA (△), JENCKEL and H. ILLERS (◐), BECKER (◑) and YAMAMOTO and WADA (◓)

dispersion around 170° K at 10^3 cps. A "third" dispersion region at low temperatures is indicated in the mechanical data of MAXWELL which covers a frequency range from about 10^{-3} to 10^2 cps including temperatures from 253 to 353° K. Using the data collected in Fig. 4, activation energies of 21 kcal/mole for the β process and 7 kcal/mole for the γ process are calculated assuming these transitions to be rate processes.

The β dispersion has been attributed to the rotation of the methoxycarbonyl side group in poly-(methyl methacrylate) by HOFF, ROBINSON and WILLBOURN. HEIJBOER has proposed that for this dispersion to be apparent some steric hindrance along the backbone chain, such as provided by the methyl group, must be present. Evidence for this was found from the results of dynamic mechanical measurements of a large number of copolymers of methyl methacrylate and other selected monomers. For a series of copolymers of methyl methacrylate and methyl acrylate the β peak gradually decreases in height in the 0.5 to 10^3 cps range at a constant temperature of 293° K, as the amount of methyl acrylate in the copolymer is increased. The damping maximum could also be depressed if rotation is overly blocked such as by copolymerization with: (1) a highly polar monomer, N-methyl methacrylamide, (2) a monomer containing a bulky side group such as cyclohexyl methacrylate or phenyl methacrylate, (3) a monomer combining polarity and steric hindrance removal, for example acrylic acid or methacrylonitrile, or (4) a crosslinking agent like ethylene dimethylacrylate. The addition of a plasticizer such as dibutyl phthalate gives an increase to the height of the maximum indicating that a larger number of methyl carbonyl groups are participating.

DEUTSCH, HOFF and REDDISH have found that upon substitution of an α chloro group for the α methyl group in poly-(methyl methacrylate) a β damping peak is still apparent although it is shifted to higher temperatures; an activation energy for the dispersion for this polymer has been reported as 30 kcal/mole. HOFF, ROBINSON and WILLBOURN have shown that the β dispersion also is exhibited by a number of poly-(methacrylate esters) and poly-(α chloroacrylate esters) being absent only in cases where the main softening region occurs at lower temperatures, thereby obscuring the β region; some of the data presented by these authors are reproduced in Fig. 5. Evidence that the main softening region can include a secondary dispersion as well has been given recently by FERRY et al. for poly-(ethyl methacrylate) and CHILD and FERRY for poly(n-butyl methacrylate) using a reduced variable analysis.

The γ dispersion region in poly-(methyl methacrylate) is still in doubt and therefore little discussion has been presented concerning a molecular explanation for it except the suggestion by YAMAMOTO and WADA that it may be connected with some type of rotation of the side

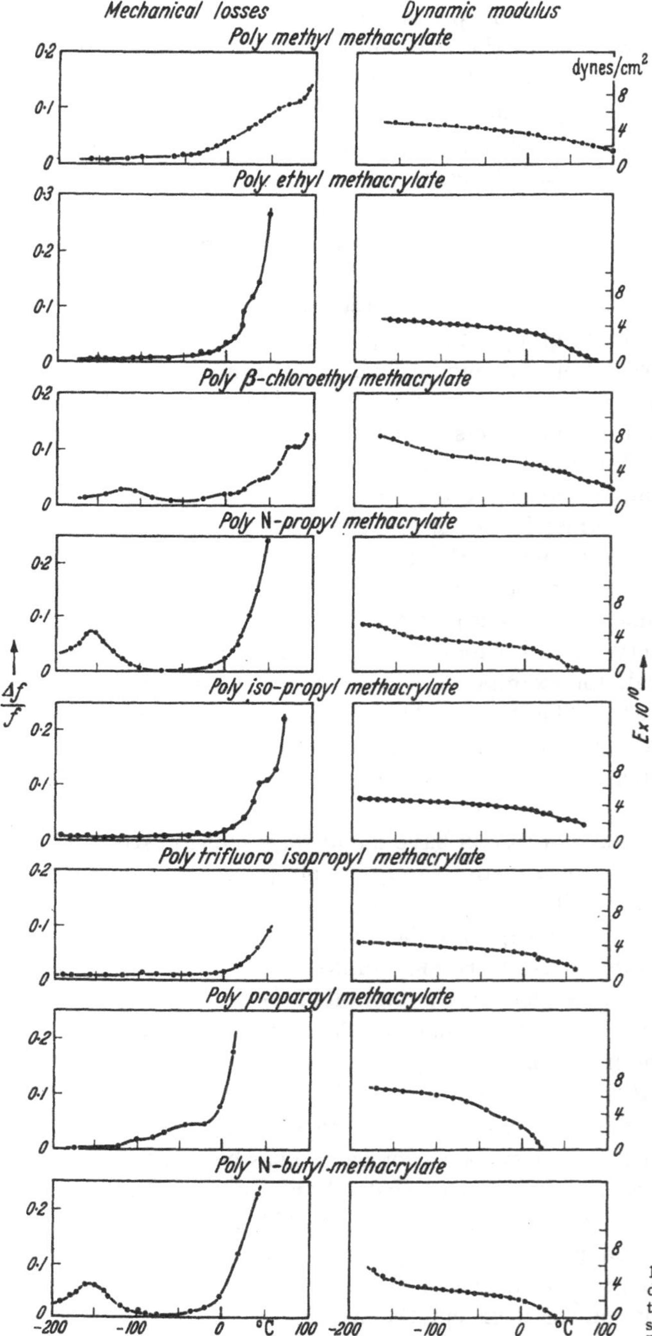

Fig. 5. Mechanical loss and dynamic modulus as functions of temperature for some polymethacrylates. Data of HOFF, ROBINSON and WILLBOURN

group. These authors also have reported that the addition of water to this polymer results in a shift of this dispersion to higher temperatures at a constant frequency, attributing it to a loss of free volume in the polymer.

FUKADA has derived curves of the loss tangent and the elastic modulus versus frequency for poly-(methyl methacrylate) from creep measurements which agree well with the experimental values over a frequency range of 10^{-2} to 10^4 cps.

HOFF, ROBINSON and WILLBOURN have reported a low temperature γ-peak at around 120° K in the 100—500 cps frequency range for poly-(n-propyl-, -(n-butyl- and -(sec-butyl-methacrylates) and -α chloroacrylates); their mechanical loss-temperature curves for various polymethacrylate esters are given in Fig. 5. This transition was attributed to the fact that more than one spatial configuration of the side-chain can exist. The fact that for the poly-(β chloroethyl-, -(neopentyl carbinyl-, and -(stearyl-methacrylates) this low temperature dispersion is found about 20—30° higher was explained as being due to polar forces, steric hindrance and inter-crystalline restraints in the side chains, respectively. Activation energies of 3—6 kcal/mole for this process were derived from overtone measurements. Poly-(methacrylate esters) which do not show this low temperature transition are the methyl, ethyl, isopropyl, trifluoroisopropyl, propargyl, tert-butyl, phenyl, pinacolyl and glycol dimethacrylate. It was also absent in poly-(methyl-, -(ethyl-, and -(isopropyl-α chloroacrylates). The lack of a low temperature peak for these materials was attributed either to side chain shortness or to inflexibility. Since polymethyl methacrylate does not have the requisite number of flexible chain atoms beyond the $-\overset{\overset{O}{/\!/}}{C}-O-$ group in the side-branch, the so-called γ-process reported for that material should *not* be identified with the γ-process exhibited by the higher esters.

For poly-(cyclohexyl methacrylate) and poly-(cyclohexyl α-chloroacrylate) a low temperature dispersion centered around approximately 240° K has been found by HOFF, ROBINSON and WILLBOURN in the 100—500 cps region. HEIJBOER has also reported the same peak for poly-(cyclohexyl acrylate) as well as poly-(cyclohexyl methacrylate). Data of HEIJBOER at four frequencies —0.4, 2, 200 and 10^3 cps — for the damping in poly-(cyclohexyl methacrylate) as a function of temperature are given in Fig. 6. HEIJBOER found that neither poly-(cyclopentyl methacrylate) nor poly-(phenyl methacrylate) exhibits this peak whereas poly-(methyl methacrylate) plasticized with cyclohexyl methacrylate does exhibit a damping maximum in this region. HOFF, ROBINSON and WILLBOURN concluded that this dispersion was caused by a "chair" to "boat" rearrangement of the cyclohexane ring. This assignment was based partially on the fact that the activation energy of 11 kcal/mole found for this dispersion was close to the 14 kcal/mole given by BECKETT, PITZER and SPITZER for this rearrangement of cyclohexane. On the other hand, HEIJBOER has shown that extrapolation of his results to higher frequencies indicates the identification of this damping maximum for poly-(cyclohexyl methacrylate) with the maximum for low molecular weight cyclohexyl derivatives at ultrasonic frequencies as reported by KARPOVICH. This dispersion was shown by KARPOVICH to be related to the transition of the cyclohexyl group from one "chair" configuration to another; cyclohexane alone does not show this dispersion. HEIJBOER also studied the effect of a paramethyl substituent in the cyclohexyl ring, finding it to increase the steric hindrance in the ring, decreasing the damping maxima but leaving the temperature-frequency diagram essentially unchanged.

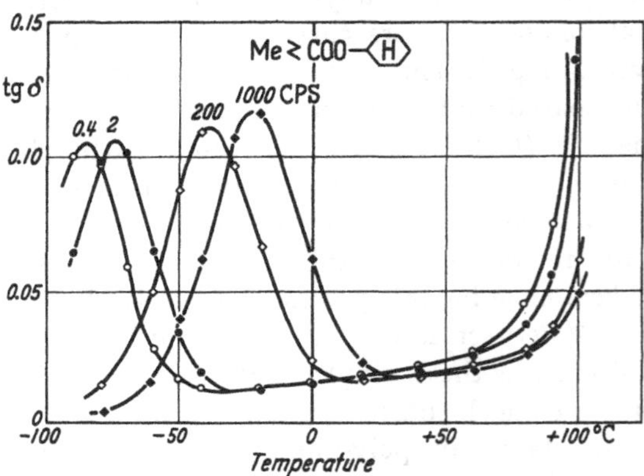

Fig. 6. Mechanical damping of poly-(cyclohexyl methacrylate) as a function of temperature for different frequencies. Data of HEIJBOER

Data on poly-(methyl-, -(ethyl- and -(n-butyl acrylate) have been reported by SCHMIEDER and WOLF (1953) at around 10 cps and by THURN and WOLF at 2×10^6 cps. Poly-(methyl acrylate) has been studied by IWAYANAGI (1) at three frequencies; he found distinct differences in the location of the primary damping peak for two samples prepared in different ways.

The more readily discernible damping maxima reported for the polyacrylates are listed in Table 1. The methyl ester is found to exhibit a secondary peak at low frequencies but none was evident from the data taken at 2×10^6 cps. IWAYANAGI's (1) investigation, carried out at both high and low frequencies, was not taken to low enough temperatures to clarify the other findings. Also, SCHMIEDER and WOLF (1953) do not give the mechanical loss v. temperature plot for this material, only listing the temperature maximum at the frequency of the investigation. At high frequencies (2×10^6 cps) the ethyl and butyl esters showed two secondary dispersion regions; at about 10 cps these esters showed β dispersions, but only for the butyl ester was the γ dispersion very marked. The temperatures at which the β dispersions are found appear in all cases to be lower than that reported for the like polymethacrylate and

Table 1. *Dispersion Regions for Some Acrylates, Vinyl Ethers and Vinyl Esters*[1]

Polymer	Dispersion						Reference
	γ		β		α		
	f cps	*T* ° K	*f* cps	*T* ° K	*f* cps	*T* ° K	
A. Polyacrylates							
methyl	—	—	—	—	.5	282—293	Iwayanagi (1)
methyl	X	X	9.5	213	1.2	298	Schmieder and Wolf (1953)
methyl	—	—	—	—	300	296—306	Iwayanagi (1)
methyl	—	—	—	—	10^3	298—311	Iwayanagi
methyl	?	?	X	X	2×10^6	339	Thurn and Wolf
ethyl	11	173(?)	9.5	223	1.7	268	Schmieder and Wolf (1953)
ethyl	2×10^6	191	2×10^6	253	2×10^6	305	Thurn and Wolf
butyl	11	133	9	193	1.25	239	Schmieder and Wolf (1953)
butyl	2×10^6	205	2×10^6	240	2×10^6	270	Thurn and Wolf
B. Poly(Vinyl Ethers)							
methyl	X	X	X	X	27.2	263	Schmieder and Wolf (1953)
methyl	2×10^6	200	2×10^6	224	2×10^6	302	Thurn and Wolf
ethyl	8	173(?)	7.5	200(?)	1.1	256	Schmieder and Wolf (1953)
ethyl	2×10^6	185	2×10^6	233	2×10^6	299	Thurn and Wolf
propyl	5.5	113	4.2	200(?)	1.1	246	Schmieder and Wolf (1953)
propyl	2×10^6	193	2×10^6	243	2×10^6	290	Thurn and Wolf
n-butyl	10.4	123	8	200(?)	0.8	241	Schmieder and Wolf (1953)
n-butyl	2×10^6	197	2×10^6	241	2×10^6	278	Thurn and Wolf
isobutyl	11.8	123	X	X	1.2	272	Schmieder and Wolf (1953)
isobutyl	2×10^6	?	2×10^6	221	2×10^6	321	Thurn and Wolf
tert butyl	X	X	X	X	1.7	356	Schmieder and Wolf (1953)
C. Poly(Vinyl Esters)							
acetate	11.7	173	10.7	243	1.9	306	Schmieder and Wolf (1953)
acetate	2×10^6	188	2×10^6	263	2×10^6	366	Thurn and Wolf
propionate	X	X	5.4	233	1.5	285	Schmieder and Wolf (1953)
propionate	?	<170(?)	2×10^6	245 (288)	2×10^6	327	Thurn and Wolf

[1] X signifies that no dispersion was reported although the temperature region expected to hold the dispersion in question was traversed. A dash means that the temperature range did not extend to a point where that dispersion is expected to be found. A question mark means that there is only a possibility that a peak exists.

poly-(α-chloro-acrylate) esters by 50 to 100°. If these dispersions are all due to motion involving the carboxyl group, this difference must be caused by the removal of the α substituted CH_3 or Cl group. From Heijboer's data for the (methyl methacrylate)-(methyl acrylate) copolymers it is not certain whether the β peak at high methyl acrylate amounts disappears completely, shifts to lower temperatures, or becomes obscured by the α transition. Iwayanagi (2) has argued that the β transition merges with the primary dispersion in poly-(methyl acrylate) at most frequencies. The γ dispersion for the n-butyl ester may be brought about by the same mechanism responsible for the γ-peak in the polymethacrylate and poly-(α-chloro acrylates). The γ-peak found at high frequencies for the ethyl ester cannot be explained by that mechanism.

Activation energies for the β process for the ethyl and n-butyl ester are calculated from the data given in Table 1 to be approximately 45 and 21 kcal/mole respectively, and for the γ process for the n-butyl ester 9 kcal/mole.

Jenckel and Herwig have studied the dynamic mechanical properties at approximately .1 cps of a number of (methyl acrylate)-styrene copolymers over a temperature range from 220 to 420° K. This series was found to exhibit only one primary peak at a temperature which depended on the amount of the two components in the copolymer.

4. Poly-(Vinyl Ethers) and Poly-(Vinyl Esters)

Wolf and coworkers have reported data for a number of poly-(vinyl ethers) and for two poly-(vinyl esters). Damping and dynamic modulus versus temperature data for the poly-(n-butyl-, -(isobutyl-, and -(t-butyl-vinyl ethers) found by Schmieder and Wolf (1953) at about 1—10 cps are given in Fig. 7; results from sound attentuation measurement by Thurn and Wolf at 2×10^6 cps for poly-(vinyl acetate) are given in Fig. 8. Temperatures of the loss maxima found for these polymers are listed in Table 1.

There definitely appears to be a low temperature (γ) peak for the poly-(propyl-, -(n-butyl- and possibly the -(isobutyl ethers) at approximately 120° K, the temperature where a peak is found for the higher poly-(methacrylate- and poly-(α chloroacrylate esters). Using the data from the two investigations for these three polyvinyl ethers, an activation energy of approximately 7 kcal/mole is found for this low temperature process. For poly-(vinyl methyl ether) and possibly the poly-(vinyl ether) no definite γ process is found at low frequencies ($\sim$ 10 cps) but peaks are reported using high frequencies as was found by Yamamoto and Wada to be the case for poly-(methyl methacrylate). If this dispersion in the ethers with the longer side chains is due to the availaility of alternate spatial configurations, the etherical oxygen should also be counted; it would be expected that the poly-(vinyl ethyl ether) would show this dispersion whereas for the polymethacrylates the propyl ester was the lowest straight chain ester to exhibit this peak.

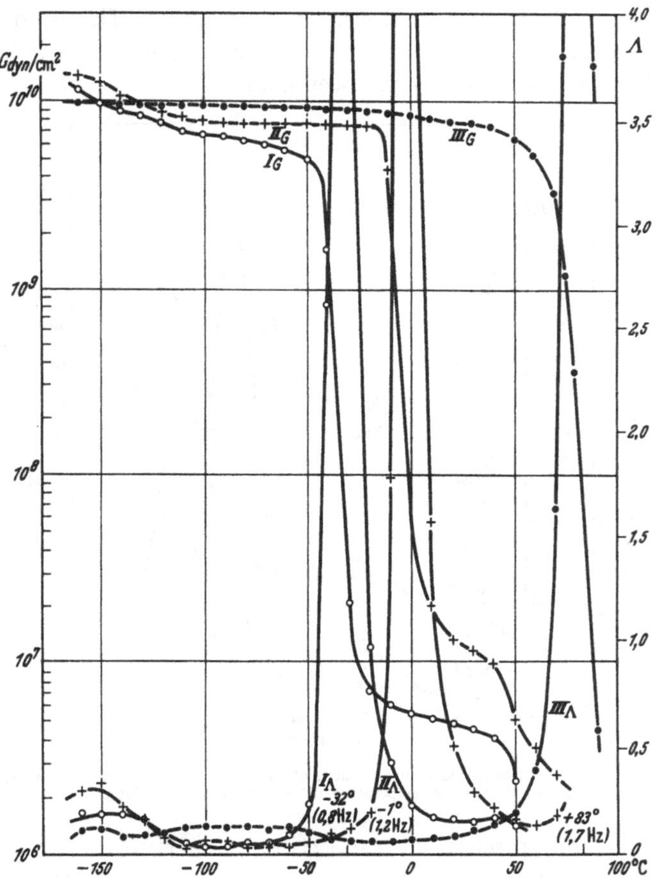

Fig. 7. Shear modulus (G) and damping (Δ) as a function of temperature for isomeric polyvinyl ethers. I. poly-(vinyl n-butyl ether); II. poly-(vinyl isobutyl ether); III. poly-(vinyl ter-butyl ether). Data of Schmieder and Wolf (1953)

Although a β process is apparent at ultrasonic frequencies for the poly-(vinyl ethers), there is only a slight indication that one occurs at lower frequencies ($\sim$ 10 cps) unobscured by the main softening process.

Poly-(vinyl acetate) clearly exhibits a γ peak at about 180° K (see Fig. 8); poly-(vinyl propionate) does not show a γ peak at low frequencies, a small rise in the damping being noticeable around 170° K at high frequencies. β dispersions at around 250° K are found for both polymers. Activation energies of approximately 40 kcal/mole for the γ process for poly-(vinyl acetate) and 80 and 120 kcal/mole for the β process in the acetate ester and the propionate ester, respectively, are calculated from the data listed in Table 1. These activation energies are much higher than those calculated for the secondary process in the poly-(acrylic esters) and poly-(vinyl ethers) given above. The β and γ dispersions are most likely associated with motions in the side branches but no definite assignments have been made.

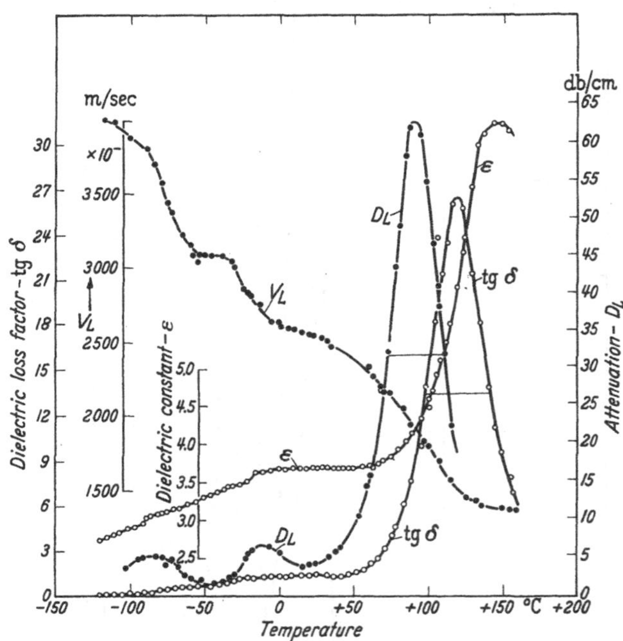

Fig. 8. Polyvinyl acetate. Dielectric loss factor, tan δ, attentuation D_L, dielectric constant, ε, and sound velocity, V_L, as a function of temperature at 2×10^6 cps. Data of THURN and WOLF

The dynamic mechanical properties of mixtures of poly-(vinyl acetate) and poly-(methyl methacrylate) have been measured by JENCKEL and HERWIG; two peaks, one characteristic of each homopolymer, were discernible for these mixtures. The same was found by JENCKEL and HERWIG for a mixture of two (methyl acrylate)-styrene copolymers.

5. Other Amorphous Polymers

The dynamic mechanical properties at low frequencies (~ 0.1 to 10 cps) of poly-(vinyl carbazole) and poly-(vinyl pyrrolidone) have been studied by SCHMIEDER and WOLF, the former showing a secondary peak at —80° C, the latter exhibiting numerous peaks below the main softening region.

WADA and YAMAMOTO first reported a secondary dispersion for a phenolic resin at ultrasonic frequencies and then in a second publication by YAMAMOTO and WADA attributed this to resonances in their instrument.

IV. Partially Crystalline Polymers

This section will be principally devoted to polymers which are partially crystalline in nature, covering materials from poly-(vinyl chloride) which is only slightly crystalline to "low pressure" polyethylene from which single crystals have been obtained by TILL. Generally, however, these materials exhibit similar dynamic mechanical properties showing in a number of cases two definite transition regions, reflected as relatively sharp drops in the dynamic modulus and the occurrence of maxima in the mechanical loss, as the temperature is increased. The crystalline to amorphous ratio of the material determines the relative magnitude of these two dispersions. Generally, the dispersion occurring at the highest temperature reflects the crystalline melting process and usually occurs somewhat below the usual crystalline melting point. Above the temperature of this dispersion region the material is rubbery or liquid-like. Below this dispersion region the next principal peak in the loss and the associated drop in modulus usually is due to a transformation in the amorphous areas of the polymer as the result of the disrupting, due to thermal agitation, of van der Waals' forces, polar forces, or even hydrogen bonds between chains. Other secondary dispersions frequently occur for partially crystalline polymers and have been attributed to crystalline phase transitions, and side chain motion or limited main chain movements in the amorphous regions.

As with the amorphous materials only a relatively few systematic studies have been carried out concerning the effects of variables, such as water content, thermal history, and per cent crystallinity, on either the primary or secondary dispersions. In addition, the physical characteristics of the material being studied, such as per cent crystallinity, density, branching content, and molecular weight, are usually not given, leading again to the difficulties outlined previously in the first part of Section III of this paper.

1. Polyethylene

Numerous investigations of the dynamic mechanical behavior of polyethylene have been reported in the recent literature. Three dispersion regions have definitely been found, these being frequency dependent but occurring in the vicinity of 170° K (γ), 280° K (β) and 350° K (α); a fourth loss region implicit in the data of NIELSEN has been reported at $\sim 400°$ K by SCHMIEDER and WOLF (1953). In one investigation reported by SAUER, FUSCHILLO, DEELEY and WOODWARD the dynamic elastic modulus in the audiofrequency region was studied as a function of temperature from 4° K up to the melting region; no dispersions in the modulus were discernible below the 80° K limit of previous investigations.

The approximate temperature positions at various frequencies for the α, β, and γ peaks are given in Table 2; data for both unbranched

Table 2. *Location of Dispersion Regions for Polyethylene*

Dispersion Region						Reference
γ		β		α		
f cps	T ° K	f cps	T ° K	f cps	T ° K	
A. "High Pressure" Polyethylene						
1.25	140	0.3	268	0.3	340	HELLWEGE, KAISER and KUPHAL
8.6	166	4.1	268	1.2	327	SCHMIEDER and WOLF (1953)
324	158	150	253	39	333	OAKES and ROBINSON
1.2×10^3	165	540	265	150	355	KLINE, SAUER and WOODWARD
1.15×10^3	165	520	280	200	360	KLINE, SAUER and WOODWARD
1.9×10^4	$\leqq 200$	6000	320	600	385	BUTTA
4×10^4	180	4×10^4	275	4×10^4	$\geqq 360$	MIKHAILOV and SOLOVÉV
1×10^5	$\leqq 190$	1×10^5	283	1×10^5	$\geqq 320$	YAMAMOTO and WADA
1×10^5	200	1×10^5	285	—	—	MIKHAILOV and SOLOVÉV
5×10^5	205	5×10^5	285	—	—	KABIN
2×10^6	210	2×10^6	295	2×10^6	360	THURN
2×10^6	210	2×10^6	300	—	—	KABIN
—	—	5×10^6	310	—	—	KABIN
B. "Low Pressure" Polyethylene and Polymethylene						
1.25	153	X	X	0.3	373	HELLWEGE, KAISER and KUPHAL
10	173	8	273	0.2	368	WOLF and SCHMIEDER
840 and 1.57×10^3	175	1.0×10^3	295	$\leqq 460$	$\geqq 380$	KLINE, SAUER and WOODWARD
1.63×10^3	177	(920)[1]	(325)[1]	(300)[1]	(400)[1]	SAUER, FUSCHILLO, DEELEY and WOODWARD
—	—	—	—	3000	420	BUTTA

("low pressure" polyethylene and polymethylene) and branched ("high pressure" polyethylene) materials are recorded. An approximate activation energy of 12 kcal/mole for the γ-process in "high pressure" polyethylene can be calculated from data gathered by HELLWEGE, KAISER and KUPHAL, by OAKES and ROBINSON, by KLINE, SAUER and

[1] Unpublished results of C. W. DEELEY.

WOODWARD, by THURN, and by KABIN over the frequency range from ~ 1 cps to 2×10^6 cps. An activation energy for the γ-process in "low pressure" polyethylene is estimated at 15 kcal/mole from the data given by SAUER, FUSCHILLO, DEELEY and WOODWARD at 1630 cps and HELLWEGE, KAISER and KUPHAL at 1.25 cps. Due to the scatter when data of various workers (see Table 2A) are plotted, an activation energy of 20—100 kcal/mole can only be estimated for the β peak for "high pressure" polyethylene; for the frequency range of 5×10^5 to 5×10^6 cps KABIN reports a value of 16 kcal/mole. Since the various investigators may very well be working with polyethylenes of different branching degree, crystallinities, and thermal histories, a large variance is not surprising. In addition to the effects of branching, changes in the dynamic mechanical properties due to orientation, high energy irradiation and chemical modification of "high pressure" polyethylene have also been studied and will be discussed below.

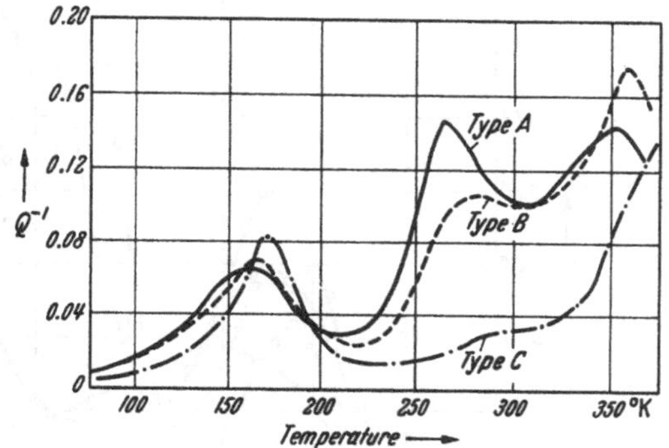

Fig. 9. Dependence of damping as a function of temperature on branching for polyethylene. Type A: 3.2 CH_3's/100 CH_2; Type B: 1.6 CH_3's/100 CH_2; Type C: < 0.1 CH_3's/100 CH_2. Data of KLINE, SAUER and WOODWARD

The effect of branching on the damping as a function of temperature at audiofrequencies is readily demonstrated by the data of KLINE, SAUER and WOODWARD given in Fig. 9. All three of the dispersion regions occurring below 380° K are affected to some extent as the amount of branching is changed. With a decrease in branching the α peak shifts to higher temperature, the room temperature peak (β) decreases in height, essentially disappearing in polyethylene containing no branches, and shifts to higher temperatures, and the low temperature region (γ) sharpens and shifts to higher temperatures.

The effects of irradiation on the dynamic modulus, sound velocity and damping of polyethylene have been studied. It has been reported by CHARLESBY, by LAWTON, BUECHE, and BALWIT, and by CHIPIRO that high energy irradiation causes crosslinking of polyethylene. CHARLESBY and ROSS first reported that irradiation led eventually to the loss of crystallinity, giving a highly rigid glass-like material at high irradiation doses. WOODWARD, DEELEY, KLINE and SAUER found that the modulus around room temperature up to crosslinking of about 20% definitely depends on the thermal history of the sample during and after irradiation, definite differences between their data and that of CHARLESBY and

HANCOCK and of BACCAREDDA and BUTTA being noted. Above the crystalline melting point, samples having received doses leading to from about 0.5 to 20% crosslinking exhibit rubber-like behavior, that is, the modulus is found to increase with temperature as shown by CHARLESBY and HANCOCK and DEELEY, KLINE, SAUER and WOODWARD. At doses giving more than 20% crosslinking, the rigidity of the specimen is too great for it to exhibit rubber-like behavior. The damping as a function

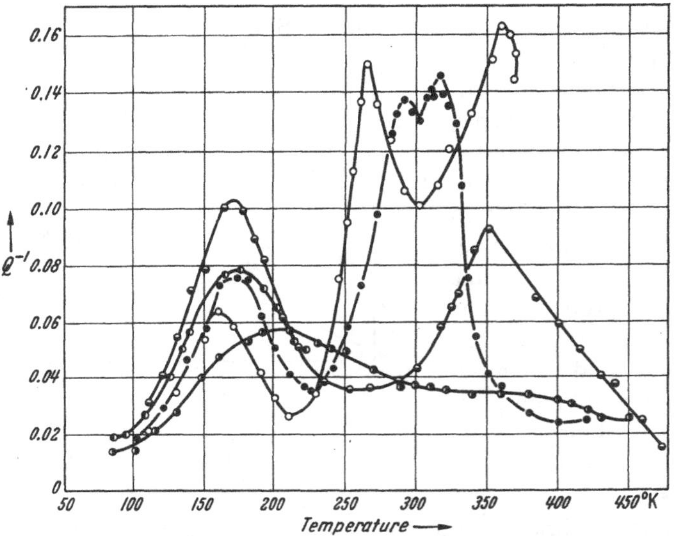

Fig. 10. Mechanical damping as a function of temperature for "high pressure" polyethylene irradiated in a reactor to dosages of: 0 nvt (○), 0.6×10^{18} nvt (●), 2.8×10^{18} nvt (◒), 5.5×10^{18} nvt (◐), and 8.3×10^{18} nvt (◑). Data of DEELEY, KLINE, SAUER and WOODWARD (note: the nvt value is the integrated thermal neutron flux in neutrons/cm²)

of temperature for polyethylene irradiated to various dosages is shown in Fig. 10 where some of the data reported by DEELEY, KLINE, SAUER and WOODWARD using a resonance technique in the 100—2000 cps range have been put on a composite graph. It is readily apparent that irradiation brings about a decrease in the α dispersion, a decrease in the β dispersion accompanied by a shift to higher temperatures, and at first an increase and then a decrease in the γ-peak height with a shift to higher temperatures taking place. The decrease in the α dispersion parallels the loss of crystallinity in the material; the changes in the β peak appear to be brought about by the greater rigidity of the irradiation cross-linked systems. The initial increase in the γ-peak coincides with the increase in the amorphous material upon crystallite destruction, the decrease in the height of this peak at higher dosages probably being caused by the hampering of motion in the amorphous areas brought about by the radiation-induced crosslinking.

HELLWEGE, KAISER and KUPHAL have found that drawing of polyethylene brings about noticeable changes in the damping, the β region at 268° K increasing in height, the γ-peak remaining essentially the same and the α peak seemingly disappearing. This has been explained as due to a loss of crystallinity during drawing with a subsequent increase in the amorphous portion of the polymer. However, if this were true, it would be expected that an increase in the γ peak would also occur as was found for the irradiated samples (see Fig. 10).

SCHMIEDER and WOLF (1953) have investigated a number of chlorinated polyethylenes using a torsional pendulum in the 1 to 10 cps frequency range. The principal effects of chlorination are: (1) the lowering of the temperature position of the crystalline melting dispersion, (2) an increase in the peak height around 270° K (the β dispersion) and at higher chlorinations a linear increase in the temperature position for this peak, (3) a decrease in the γ peak height and a shift to higher temperatures and (4) the appearance of a dispersion higher in temperature than the one due to crystalline melting, thought to be associated with a rubber-liquid transition.

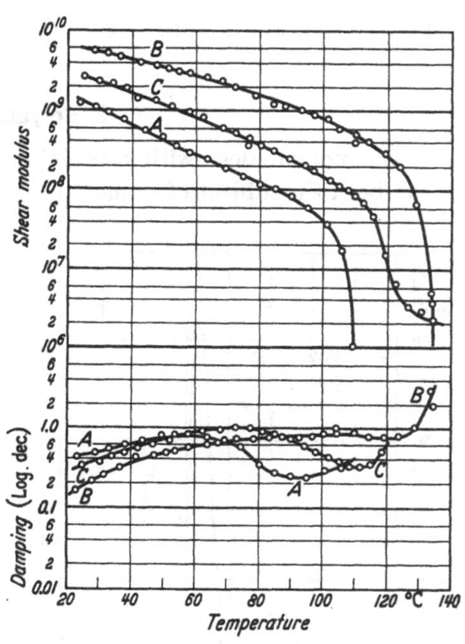

Fig. 11. Dynamic mechanical properties of polyethylenes. Data of NIELSEN

For polyethylene the α dispersion region has been attributed to the increased segmental mobility accompanying the crystalline melting process. It appears from the data of SCHMIEDER and WOLF (1953) and that of NIELSEN some of which is given in Fig. 11, that a higher temperature process occurs which may reflect a transformation from the rubbery to the liquid state. NIELSEN has shown (see Fig. 11) that both of these transitions shift to higher temperatures as the degree of branching decreases.

The β and γ dispersions are thought to be due to diffusional motion of chain segments in the amorphous regions, the β process involving the cooperation of carbon atoms at the branch points along the main backbone chain, the γ process taking place in the side branches or main chain without this cooperation. In accordance with the above interpretation the γ-process in polyethylene occurs at approximately the same temperature as the low temperature process for poly-(stearyl methacrylate).

MIKHAILOV and SOLOV'EV have attempted to relate the β peak in polyethylene to a dispersion found for a commercial paraffin sample in the 10^5 cps region at a temperature about 25° lower, concluding that the β dispersion is therefore due to some unidentified crystalline transformation. However, this assignment is not in accordance with the bulk of experimental evidence discussed above.

2. Polyamides

The polyamides, with especial emphasis on poly-(ε caprolactam), have been the subject of numerous investigations. Over the temperature range from 80° K to approximately 500° K four dispersion regions have been reported for all the polyamides investigated; these are frequency dependent but are located at about 160° K (γ), 230° K (β), 350° K (α') and just below the crystalline melting point (α). Examples of the behavior found for three polyamides in the 100 to 1500 cps region are given in Fig. 12 where the internal friction is plotted as a function of temperature. It has been found that the exact temperature position of these dispersions differs from material to material depending on the thermal history and/or crystallinity and water content of the material. Loss peak data for the polyamides are given in Table 3.

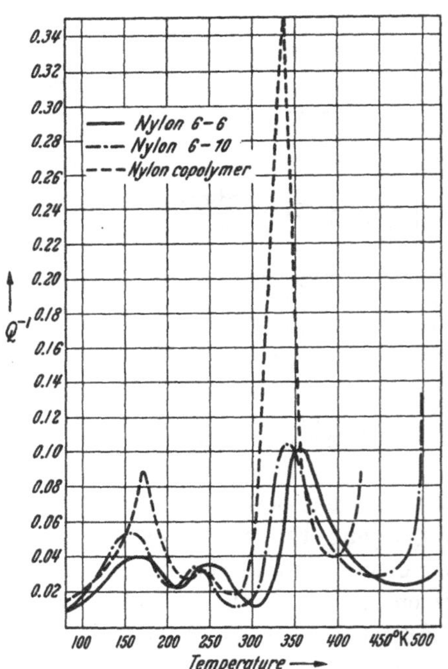

Fig. 12. Damping as a function of temperature for three polyamides. Data of WOODWARD, SAUER, DEELEY and KLINE

A number of investigations have concerned poly-(ε caprolactam) in the temperature ranges from about 200° K to 370° K where the β (~ 250° K) and the α' (~ 340° K) dispersion regions are found. Approximate activation energies calculated from the data in Table 3 are 15—20 and 95 kcals/mole for the β and α' processes, respectively.

BECKER and OBERST have studied extensively the effects of water content on the α' dispersion; some of their data are given in Fig. 13 along with data collected for some polyurethanes. A shift to lower temperatures of the α' dispersion at constant frequency as the water content is increased is apparent in Fig. 13, an effect previously reported by

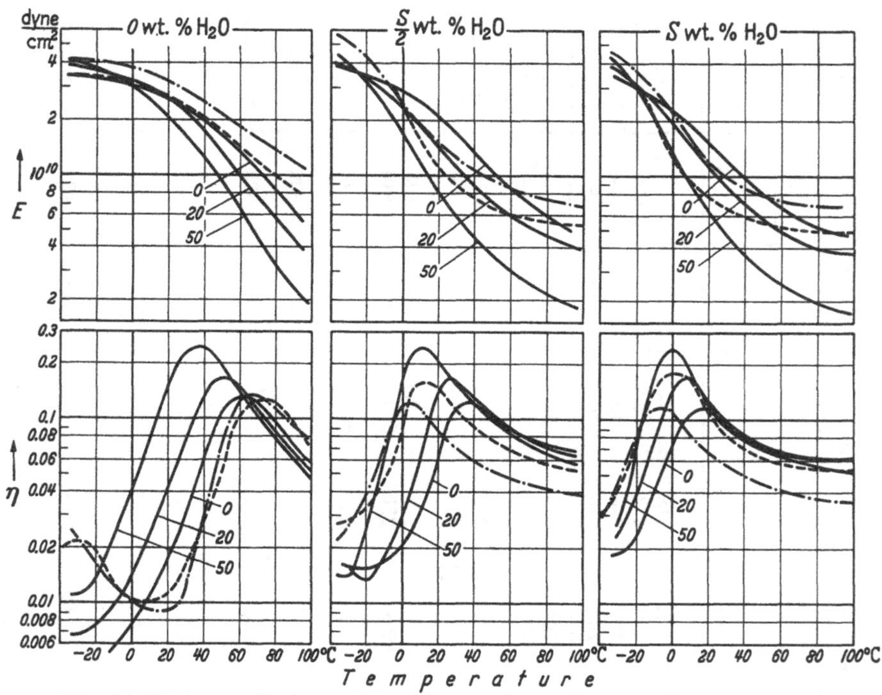

Fig. 13. Elastic modulus, E, and mechanical damping, η, as a function of temperature for different poly-(ε-amino caprolactam) and poly-urethane samples at three water contents (0, half-saturated, $S/2$, and saturated, S at a frequency of 100 cps). Data of BECKER and OBERST

WOLF and SCHMIEDER. From WOLF and SCHMIEDER and BECKER and OBERST's curves it appears that the β peak is also shifted to lower temperatures upon addition of water to a dried sample.

Fig. 13 also includes curves of BECKER and OBERST for materials at two crystallinities showing a small decrease in the peak temperature with decreasing crystallinity. This again is in agreement with WOLF and SCHMIEDER who found a shift to lower temperatures accompanied by a sharpening of the α' peak for a quenched specimen as compared to an annealed one.

The value of 310° K for the α' peak at ~ 5 cps first reported by SCHMIEDER and WOLF (1953) was found to be about 20° too low by

Table 3. *Dispersion Regions Reported for the Polyamides, Polyesters and Polyurethanes*[1]

Polymer	Damping Peaks							Reference
	γ		β		α'		α	
	f cps	T °K	f cps	T °K	f cps	T °K	T °K	
polycaprolactam (6)	8.4	153	7.1	218	—	—	—	Schmieder and Wolf (1953)
(and polycaproamide)	?	153	?	223	5.2	333	> 500	Wolf and Schmieder
	—	—	10^2	~ 243	10^2	~ 343	—	Becker and Oberst
	—	—	10^2	< 240	10^2	~ 338	—	Becker and Oberst
	—	—	1.66×10^2	~ 230	10^2	~ 350	—	Kawaguchi
	—	—	10^3	~ 258	10^3	353	—	Becker and Oberst
	10^5	$\geqq 200$	10^5	$\geqq 290$	—	—	—	Yamamoto and Wada
	—	—	1.46×10^6	305	—	—	—	Wada and Yamamoto
polycapryllactam (8)	10	143	8.1	223	5.2	322	> 470	Schmieder and Wolf (1953)
polyaminoundecanoic acid (11)	8.3	158	6.8	218	3.8	329	> 360	Schmieder and Wolf (1953)
polyhexamethylene	8	153	6.4	223	3.4	338	> 525	Schmieder and Wolf (1953)
adipamide (6—6)	1.29×10^3	165	1.07×10^3	250	660	350	> 540	Woodward, Sauer, Deeley and Kline
	1.25×10^3	175	X	X	700	365	> 540	Deeley, Woodward and Sauer
polyhexamethylene pimelamide (6—7)	7.3	158	6.0	223	3.7	331	> 490	Schmieder and Wolf (1953)
polyhexamethylene suberamide (6—8)	6.8	153	5.4	223	3.1	337	> 505	Schmieder and Wolf (1953)
polyhexamethylene	6.5	143	5.5	213	4.0	333	> 495	Schmieder and Wolf (1953)
sebacamide (6—10)	1520	155	1020	240	660	340	> 500	Woodward, Sauer, Deeley and Kline
polydecane-dicarboxylic acid hexamethylene diamine (6—12)	6.2	153	5.3	223	3.8	329	> 385	Schmieder and Wolf (1953)
polyhexamethylene methyl pimelamide	8.5	153	7.0	215	4.1	323	433	Schmieder and Wolf (1953)

polyethylene sebacamide (2—10)	9.8	153	8	218	5.5	338	523	SCHMIEDER and WOLF (1953)
polyethylene terphthalate	—	—	?	220	6	370	> 530	WOLF and SCHMIEDER
	—	—	1	230	1	370	—	THOMPSON and WOODS
	X	X	1030	260	—	—	—	DEELEY[2]
	—	—	10^4	310	10^4	410	—	THOMPSON and WOODS
polyurethane from hexane disocyanate + 1,4-butane-diol	11	147	?	~210	3.6	321	> 450	WOLF and SCHMIEDER
	—	—	—	—	100	~330	—	BECKER and OBERST
	—	—	—	—	0.5	307	—	JENCKEL

[1] Dashes and X's signify the same as in Table 1; the question marks in the frequency columns in this table indicate that the frequency for which a peak is apparent was not given.

[2] Unpublished results.

WOLF and SCHMIEDER due to the presence of residual monomer; WOLF and SCHMIEDER also report a small shift in the β dispersion to lower temperatures when residual monomer is present.

Only YAMAMOTO and WADA and SCHMIEDER and WOLF (1953) have carried their investigations to low enough temperatures at the measuring frequency used to show the existence of a γ peak ($\sim$ 175° K) in poly-(ε caprolactam). Thermal history appears to effect this peak although the only example of this has been given by WOLF and SCHMIEDER. The effect of water or other low molecular weight compounds has not been reported.

As is apparent from Table 3, SCHMIEDER and WOLF (1953) have investigated a large number of other polyamides over a broad temperature range at low frequencies. Poly-(hexamethylene adipamide) and poly-(hexamethylene sebacamide) have also been studied in the 100 to 1000 cps range by WOODWARD, SAUER, DEELEY and KLINE. From the data given in Table 3 activation energies expressed in kcals/mole for the three lower temperature processes in poly-(hexamethylene adipamide) and poly-(hexamethylene sebacamide), respectively, are calculated to be 22 and 20 for the γ-process, 21 and 20 for the β process and 50—75 and 170 for the α' process. Since these values are derived from data at essentially only two frequencies, they are considered to give orders of magnitude only; the two figures for the α' peak for poly-(hexamethylene adipamide) were calculated using data for samples studied in the 1000 cps range with different thermal histories.

Upon investigating essentially amorphous polyamides, for example a copolymer of ε caprolactam, hexamethylene diamine, adipic acid and sebacic acid, all the dispersions are seen to change by a marked degree.

This is quite apparent from data of WOODWARD, SAUER, DEELEY and KLINE given in Fig. 12. For the essentially amorphous copolymer the α ($\sim$ 500° K) and α' regions shift to lower temperatures, the α' region increases markedly in height and area, and the γ peak increases in area

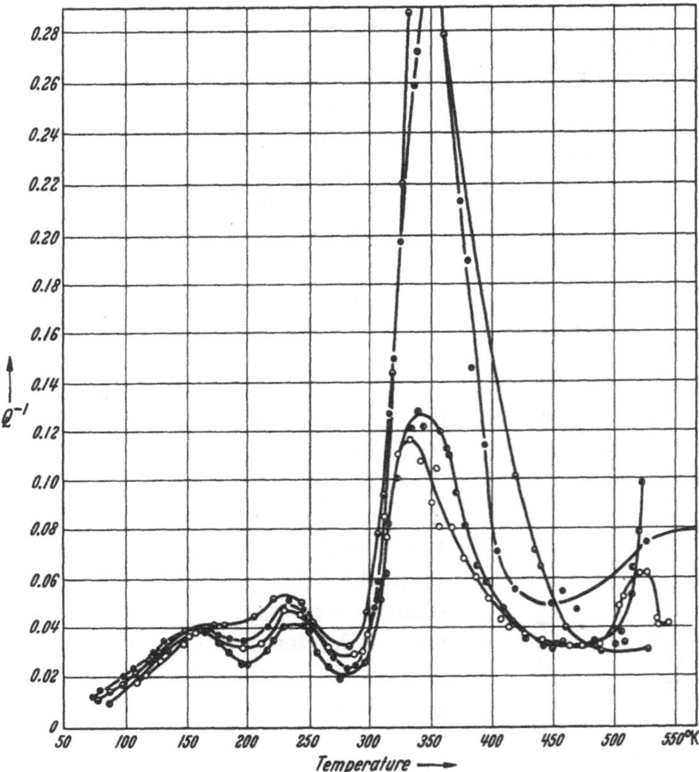

Fig. 14. Mechanical damping as a function of temperature for poly-(hexamethylene adipamide) irradiated in a reactor to dosages of: 0 nvt (◐), .3 × 10^{18} nvt (○), 2.8 × 10^{18} nvt (●), 5.5 × 10^{18} nvt (◒). Data from DEELEY, WOODWARD and SAUER (note: the nvt value is the integrated thermal neutron flux in neutrons/cm²)

as compared with these same regions in poly-(hexamethylene adipamide) or poly-(hexamethylene sebacamide).

Some effects of thermal history and/or water content on the dynamic mechanical properties of poly-(hexamethylene adipamide) have been studied. WOODWARD, SAUER, DEELEY and KLINE have shown that an increase of water content definitely shifts the α' peak to lower temperatures and that annealing treatments, which may have also brought about further drying, principally lower the β dispersion. In fact, as found by DEELEY, WOODWARD, and SAUER the β region is essentially absent if the proper treatment is used, an increase in the γ dispersion also being noted.

However, the γ dispersion does not appear to be as readily changed by thermal treatment as the β region.

The effects of high energy irradiation on the dynamic mechanical properties of poly-(hexamethylene adipamide) have been recently studied by DEELEY, WOODWARD and SAUER. The dipersion regions found to be principally affected were the α and α' ones as is shown in Fig. 14, where some of the data gathered by the above authors in the 100—1500 cps range for irradiated specimens are given on a composite graph of Q^{-1} against temperature. The decrease in the α region seen in this figure can be correlated with the loss of crystallinity due to the irradiation; the increase in the α' peak was attributed to the increase in the amount of amorphous material. However the lack of a decrease in the α' region at higher dosages and the rather minor effects of irradiation on the β and γ dispersions have not been satisfactorily explained.

The four dispersions found for all polyamide samples have been attributed to the following: the α region ($> 500°$ K) is the consequence of the melting of crystallites; the α' region (350° K) is due to the increased mobility in the amorphous regions accompanying the disrupting of hydrogen bonds in those areas; the β dispersion (230° K) reflects the onset of motion of chain segments in the amorphous regions involving the cooperation of amide groups not hydrogen bonded to other amide groups; and the γ region ($\sim 160°$ K) is caused by the cooperative motion of CH_2 groups between amide groups. It has been pointed out that the crystalline phase transition reported by BRILL for polyhexamethylene adipamide at 436° K may be reflected in the high temperature side of the α' peak.

3. Polyesters

The polyesters have been the subject of much less extensive studies than the polyamides. With the exception of some damping and modulus curves for unidentified cross-linked polyesters the main body of published material has been on poly-(ethylene terephthalate). This polymer has been investigated by WOLF and SCHMIEDER, WOODS, and THOMPSON and WOODS.

The most extensive investigation of this polymer has been published by THOMPSON and WOODS, who studied amorphous, undrawn crystalline, and oriented crystalline poly-(ethylene terephthalate) at frequencies of 10^{-3} to 10 cps and 10^4 cps. The frequency dependence of the modulus and loss angle for the unoriented crystalline material over a temperature range of 190 to 450° K is shown in Fig. 15. From the data in this figure it is apparent that at least two transitions exist for this polymer at approximately 230° K (γ) and 370° K (β); these are believed to be related to motions in the amorphous regions since similar loss maxima are found in an almost completely amorphous sample. A third loss peak occurring at still higher temperatures has been found by Wolf and Schmieder and has been designated as the dispersion due to crystalline melting.

Thompson and Woods showed that the peak height, peak width, transition temperature, and the frequency dependence of the dispersion

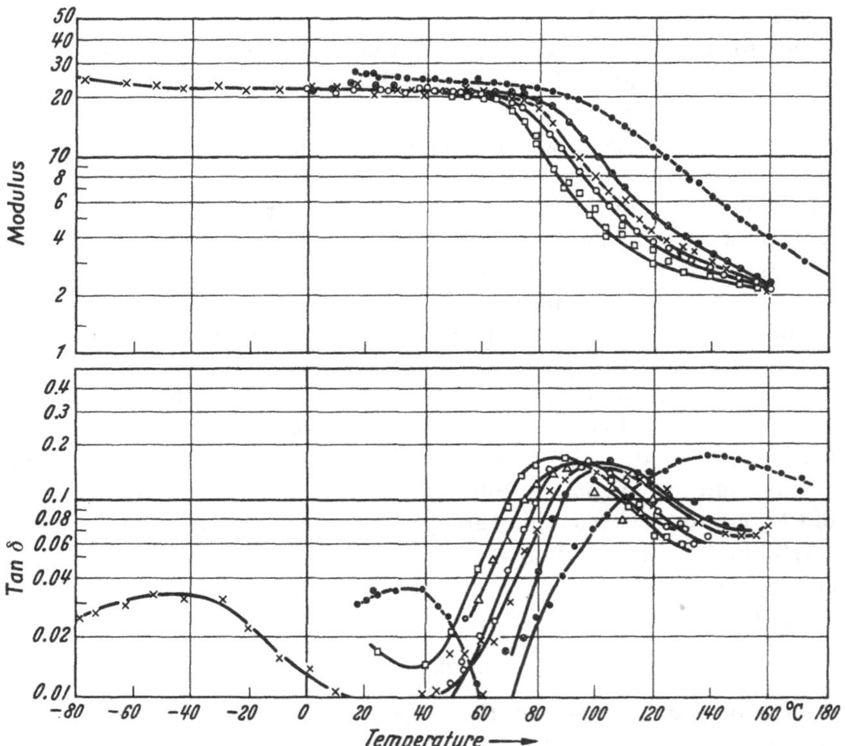

Fig. 15. Modulus and loss angle, tan δ, of unoriented crystalline poly-(ethylene terephthalate) at various frequencies as a function of temperature, 10^1 cps (●), 5.9 cps (⊗), 0.72 cps (×), 0.082 cps (○), 0.0094 cps (△), 10^{-3} cps (□). Data of Thompson and Woods

at around 370° K changed markedly as the degree of crystallinity of the material was altered. As the crystallinity is increased, going from an amorphous quenched sample to an unoriented crystalline and then to an oriented crystalline material, the peak height decreases, the peak width increases, the transition temperature increases from 350 to 400° K, and the activation energy decreases. The latter value changes from 182 kcal/mole for the amorphous material (2% crystallinity) to 97 kcal/mole for the oriented crystalline sample (65% crystallinity). Although the lower temperature peak was not studied as extensively, it was reported that it is less affected by crystallinity with only a small shift to higher temperatures of the dispersion as the crystallinity is increased. The activation energy was not greatly affected by the change, being approximately 17 kcal/mole. Some loss peak data for crystalline poly-(ethylene terephthalate) are given in Table 3.

THOMPSON and WOODS put forth three possible assignments for the β and γ dispersions in poly-(ethylene terephthalate). From considerations of the damping peaks in other partially crystalline materials, the most acceptable explanation is that the upper transition is due to rotational motion of the aromatic parts of the chain around the para-phenylene bonds whereas the lower transition is brought about by a rotational-diffusional movement of the —O—CH_2—CH_2—O— segments of the molecules.

4. Polyurethanes

Results for the polyurethane derived from hexane diisocyanate and butane-1,4 diol have been published by WOLF and SCHMIEDER, BECKER and OBERST, and JENCKEL (see Table 3). A peak at about 320° K is definitely found. A dispersion region at lower temperatures and at least one at higher temperatures are indicated by the data of WOLF and SCHMIEDER. An activation energy of approximately 40 kcals/mole can be calculated for the α' peak from the data given in Table 3. BECKER and OBERST have studied modified polyurethanes but do not detail the modifications except to indicate that the % crystallinity is changed; a shift of the 320° K peak to lower temperatures parallels the decrease in crystallinity for those materials. WOLF and SCHMIEDER and BECKER and OBERST indicate that this peak in the polyurethanes occurs at lower temperatures the greater the water content as is shown in Fig. 13, a shift also found for the polyamides. Changes in the size of the two low temperature peaks in the polyurethane investigated by WOLF and SCHMIEDER also appear to occur with the addition of water. With less certainty a similar assignment of mechanisms to the four dispersion regions as given for the polyamides appears to apply to the polyurethanes (see Section IV 2).

JENCKEL has studied the effect of drawing, coupled with thermal treatments, on the dynamic mechanical behavior of polyurethanes. For undrawn specimens the maximum at about 310° K decreased in height as the annealing temperature was increased. However, in the case of a drawn sample, just the opposite occurred. The annealing treatment afforded these samples was not described in detail but presumably the crystallinity of the undrawn specimens was increasing with a rise in the annealing temperature while it decreased for the drawn specimens due to a melting of the oriented crystallites.

5. Poly-(Vinyl Chloride) and Poly-(Vinylidene Chloride)

Poly-(vinyl chloride) is a strongly polar, slightly crystalline polymer. The crystalline regions are believed (see ALFREY, WIEDERHORN, STEIN and TOBOLSKY) to comprise but a small fraction of the total volume and the crystallites themselves are small with dimensions of the order of 100 Å. From dielectric measurements of FUOSS and others it is known

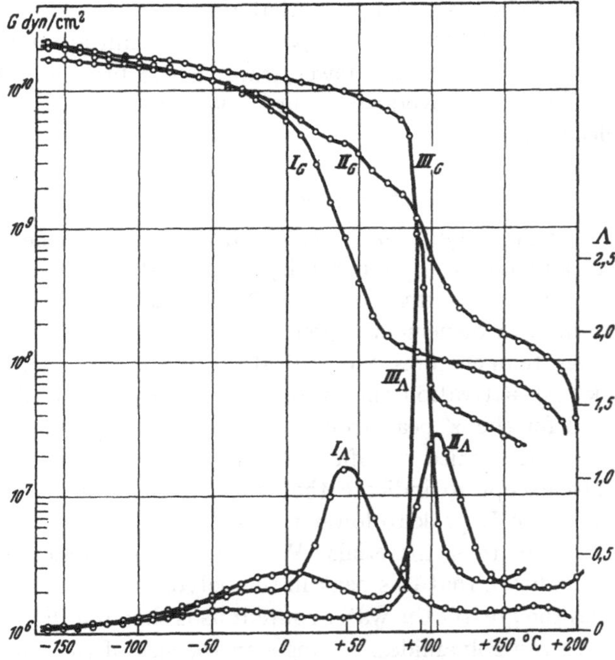

Fig. 16. Mechanical damping, Λ, and shear modulus, G, as a function of temperature for: I. poly-(vinyl fluoride); II. polytrifluoromonochloroethylene; III. poly-(vinyl chloride). Data of WOLF and SCHMIEDER

that poly-(vinyl chloride) undergoes at least two temperature transitions. This is also revealed by the dynamic mechanical data, especially that of SCHMIEDER and WOLF and SAUER and KLINE (1956) which extended over a temperature range from liquid nitrogen values to the melting range.

The test results of WOLF and SCHMIEDER are shown in Fig. 16 which also includes the measured moduli and decrement for two somewhat similar polymers, poly-(vinyl fluoride) and polytrifluoromonochloroethylene. Three transition regions are clearly evident in all materials. Various loss peak data are collected in Table 4. In poly-(vinyl chloride) the primary or α transition, during which the shear modulus drops from about 7×10^9 dyne/cm² to about 5×10^7 dyne/cm², has its damping peak at 360° K at a test frequency of 0.67 cps while the secondary or

Table 4. *Damping Peaks for Various Partially Crystalline Polymers*[1]

Polymer	γ		β		α'		α	References
	f cps	T °K	f cps	T °K	f cps	T °K	T °K	
poly-(vinyl chloride)	X	X	85	245	0.67[2]	365[2]		SCHMIEDER and WOLF (1953)
poly-(vinyl chloride)	X	X	570	273	—	—	—	SAUER and KLINE (1956)
poly-(vinyl chloride)	—	—	500	283	500	380		BECKER
poly-(vinyl chloride)	X	X	2×10^6	?	2×10^6	380		THURN
poly-(vinylidene chloride)	X	X	11	288	5.5	353	> 465	SCHMIEDER and WOLF (1953)
polytetrafluoroethylene	?	158	?	298	?	393	613	SCHULZ
polytetrafluoroethylene	12.4	203	6.3	304	2.7	403	—	SCHMIEDER and WOLF (1953)
polytetrafluoroethylene	7.5	202	4.3	306	2.0	404	560	WOLF and SCHMIEDER
polytetrafluoroethylene	730	198	690	320	400	413	—	SAUER and KLINE (1956)
polytetrafluoroethylene	5×10^5	250	—	—	—	—	—	KABIN
polytetrafluoroethylene	5×10^6	240	—	—	—	—	—	KABIN
polytrifluoromonochloro-ethylene	X	X	?	240	?	368	483	SCHULZ
polytrifluoromonochloro-ethylene	X	X	12	273	3.3	377	> 475	SCHMIEDER and WOLF (1953)
poly-(vinyl fluoride)	X	X	5.4	253	1.7	311	> 465	SCHMIEDER and WOLF (1953)
polyacrylonitrile	10	180—320	5.2	378	4.4	413		SCHMIEDER and WOLF (1953)

[1] A question mark in a frequency column signifies that the frequency was not given, one in a temperature column that a transition was not found at that given frequency. The X's and dashes signify the same as in Table 1.

[2] This dispersion is referred to as the alpha peak in the text.

β dispersion occurs below room temperature. The secondary damping peak is found at 240° K at a test frequency of 85 cps. The transverse beam tests of SAUER and KLINE (1956) place the secondary damping maximum about 30° higher but the test frequency at the peak (570 cps) is also considerably higher.

The secondary transition in the mechanical properties is also revealed by the work of BECKER who carefully explored by use of a resonance vertical vibrating reed technique a broad frequency range from 1 to 10,000 cps. Since his data, however, were taken only from 270° K to the melting range the damping only shows the secondary β peak at the frequencies at 50 cps and higher. At 500 cps BECKER reports the β peak at about 280° K and the α peak at 380° K. At a frequency of 5000 cps the reported positions are, respectively, 310 and 390° K.

The dynamic mechanical behavior at a very much higher frequency, 2×10^6 cps, has been investigated by THURN who reports measurements of attentuation and sound velocity, as well as dielectric constant, over a temperature range from 170 to 450° K. His results give no clearly discernible β peak but do show a very large α attenuation maximum at 380° K. However there are several shoulders on his attenuation-temperature curve occurring in the region of 220° K, 273—283° K and about 350° K. Unfortunately these shoulders do not seem to correlate very well with the positions of the sharper changes in dielectric constant. Other ultrasonic measurements of 10^5 cps have been reported by WADA and YAMAMOTO using a piezo-electric composite oscillator method. They report a transition temperature of 310° K for poly-(vinyl chloride) at which point they observe a change in shape of the sound velocity-temperature curve. Since they report no higher temperature transition region it would appear that this is the primary or α transition. Its unusually low value may arise from the presence of plasticizer or monomer in their specimen.

It is well known that the α peak will move to lower temperatures if plasticizer is added to pure poly-(vinyl chloride) with the decrease being essentially proportional to the amount of plasticizer present. In fact the primary glass transition may well be found below 273° K if sufficient plasticizer is present. For example, SCHMIEDER and WOLF (1952) have shown that poly-(vinyl chloride) plus 56.5% dibutyl phthalate will reduce T_α from 365 to 210° K; WALTER reports that poly-(vinyl chloride) +90% dioctyl phthalate will lower T_α to 200° K; and WURSTLIN has shown that poly-(vinyl chloride) +30% tricresyl phosphate will reduce T_α from 350 to 230° K. From dielectric measurements, HARTMANN finds that addition of plasticizer will not only reduce the α peak but may also introduce a second lower temperature peak characteristic of the behavior of the solvent rather than the polymer. For example when 16% acetophenone is added to poly-(vinyl chloride) only one dielectric dispersion is observed over the temperature range from 230 to 350° K and that is the α dispersion at 340° K. However if 33% of the same plasticizer is added the α peak falls to 298° K and an entirely new loss factor peak is found near 240° K. It has also been shown by WÜRSTLIN that in poly-(vinyl chloride) plus tricresyl phosphate a secondary maximum will arise as a result of undissolved plasticizer.

The effect of molecular weight distribution on the dynamic mechanical properties of polyvinyl chloride has been studied by BUCHDAHL and NIELSEN by a torsional pendulum method. They showed that in the temperature region from about 310 to 360° K the dispersion region was unchanged whether they tested a poly-(vinyl chloride) fraction or the unfractionated polymer. Small changes in degree of branching also had no effect but the addition of bulky end groups did introduce a secondary loss maximum below the primary one.

An activation energy for the α transition has been estimated by BECKER to be about 77 kcal/mole. Neglecting differences between specimens, one can obtain a very approximate value of 15—20 kcal/mole for the activation energy of the secondary or β transition by comparing the temperature and frequency location of this peak from the data of SCHMIEDER and WOLF, SAUER and KLINE (1956), and BECKER. The high activation energy for the α transition leads one to believe that this transition is associated with the overcoming of the dipolar and interchain cohesive forces and involves the movement of whole molecules. No definite interpretation has as yet been suggested for the occurrence of the β peak but the much lower activation energy would indicate that much smaller groups of polymer segments were involved.

Poly-(vinylidene chloride) examined by SCHMIEDER and WOLF (1953) exhibits three dispersions as given in Table 4. The lower temperature dispersion at about 290° K appears to have the largest drop in modulus with increasing temperature and therefore is probably the principal dispersion connected with the breaking up of polar bonds in the amorphous regions. The dispersion around 465° K is attributed to the crystalline to liquid transition. SCHMIEDER and WOLF (1953) have attributed both peaks in the 273 to 373° K range found for poly-(vinylidene chloride) to segmental motions in the amorphous regions, the lower temperature part being due to essentially non-stressed areas, the higher temperature side being connected with stressed parts of the chains. This explanation has been used by those authors to explain similar behavior in a number of partially crystalline polymers. That partially crystalline materials exhibit amorphous loss peaks, which are broader and occur over a wider temperature range than the similar peaks in the same material in the quenched and therefore highly amorphous state, has been demonstrated by WOODS for poly-(ethylene terephthalate) and SCHMIEDER and WOLF (1953) for some polyamides. However, one would expect that a continuous spectrum of stressed states would be evident and therefore only one, not two or more mechanical loss peaks, would be found. As suggested below for polytetrafluoroethylene, some of these dispersions may very well be due to crystalline phase transitions not reported to date.

SCHMIEDER and WOLF (1953) have also studied a number of (vinylidene chloride)-(vinyl chloride) copolymers. The crystalline melting peak of poly-(vinylidene chloride) was shown to shift to lower temperatures with increasing amounts of vinyl chloride; the two lower temperature peaks in poly-(vinylidene chloride) were found to merge into one for pure poly(-vinyl chloride), indicating that they occur in the same regions, that is, the amorphous parts of the polymer.

6. Polytetrafluoroethylene, Polytrifluoromonochloroethylene, and Poly-(Vinyl Fluoride)

Polytetrafluoroethylene has been investigated in three frequency ranges and at least four peaks, whose characteristics are listed in Table 4,

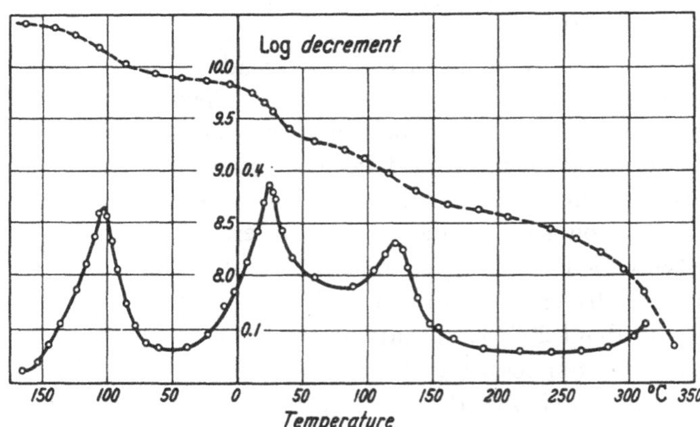

Fig. 17. Torsional modulus, G, and log decrement for polytetrafluoroethylene as a function of temperature at a frequency of 0.2 to 1 cps. Data of SCHULZ

are exhibited by this material. Data gathered over the temperature range of 80 to 590° K, at a frequency which varied with temperature from about 1 to 0.2 cps are given in Fig. 17. The peak at about 200° K has been likened to the low temperature peak in polyethylene at about 160° K and is attributed to motions of $—CF_2—$ groups following the loss of interchain binding in the amorphous areas. Indeed it has recently been shown independently by KABIN working in the megacycle range and MCCRUM in the 1—10 cps range that the area under this peak decreases when samples of increasing crystallinity are measured. KABIN reported an activation energy of 18 kcals/mole for this dispersion using his data and that of SAUER and KLINE (1955) at 730 cps. WOLF and SCHMIEDER have attributed the dispersion in the room temperature region to "stressed" amorphous regions. This assignment appears unsatisfactory for reasons discussed above. It appears more likely that this dispersion reflects the two crystalline phase transitions, attributed by PIERCE, CLARK, WHITNEY and BRYANT to the uncoiling of the polytetrafluoroethylene helix, found at 293 and 303° K. A further change in the X-ray pattern noted at elevated temperatures by the same authors may be correlated with the dispersion found around 400° K.

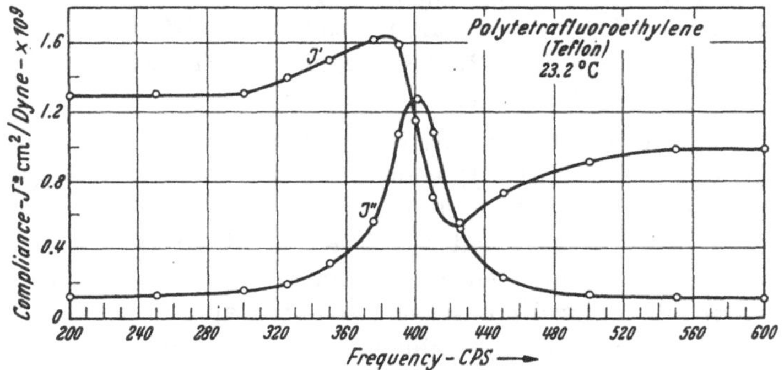

Fig. 18. Variation of complex shear compliance ($J^* = J' - i\,J''$) with frequency (linear plot) for a sample of polytetrafluoroethylene. Data of FITZGERALD

Recently measurements of the dynamic modulus and damping of polytetrafluoroethylene and two copolymers of tetrafluoroethylene and hexafluoropropylene from 4.2 to 90° K have been made by SINNOT using a torsional pendulum. No transitions were evident over this temperature range for any of these three polymers.

FITZGERALD has studied the complex shear compliance of polytetrafluoroethylene in the 298° K region over a frequency range of 100 to 5000 cps and has found a number of very sharp dispersions of the resonance rather than the relaxation type. An example of one such dispersion is seen in Fig. 18 where the real (J') and imaginary (J'') parts of the complex compliance are plotted as a function of frequency in the 200—600 cps range for a polytetrafluoroethylene sample at 296° K. Poly-(vinyl stearate) was also found to exhibit multiple resonance dispersions in this frequency range. An essentially linear polyethylene at 25.2° C showed only one resonance dispersion at 2700 cps. It was found that for polytetrafluoroethylene and poly-(vinyl stearate) the frequency at which a resonance is observed as well as its magnitude is dependent on the static stress history and thermal history of the sample. Since similar results were also found for some polycrystalline metals, it appears that these dispersions are due to stress induced motions of dislocations or other imperfections in the crystalline structure.

Polytrifluoromonochloroethylene has been investigated by SCHMIEDER and WOLF (1953), and SCHULZ using torsional pendulums. The damping peak positions found for polytrifluoromonochloroethylene are given in Table 4. The loss spectrum found for this polymer, as can be seen in Fig. 16, is somewhat similar to that obtained for poly-(vinyl chloride) but quite different from that reported for polytetrafluoroethylene, exhibiting three damping regions. The two peaks found at about 370 and 475° K are attributed to the onset of segmental motion upon breaking up of cohesive forces in first the amorphous and then the crystalline regions. It should be recalled that the principal amorphous transition in polytetrafluoroethylene is believed to occur around 200° K. No definite assignment has been made for the peak at 270° K as is also the case with the similar one found for polyvinyl chloride.

Only one investigation of poly-(vinyl fluoride) has been reported [SCHMIEDER and WOLF (1953)]; again, as in poly-(vinyl chloride) and polytrifluoromonochlorethylene, three loss peaks are found (see Fig. 16). The peaks at about 310° K and > 465° K have been given the usual amorphous and crystalline melting assignments. The dispersion in the 250° K range has received no assignment but it appears to be similar to the secondary peak in polytrifluoromonochloroethylene and poly-(vinyl chloride). It should be pointed out that the two fluoropolymers showing a greater area under this peak are also more highly crystalline than poly-(vinyl chloride).

7. Other Partially Crystalline Polymers

Polyacrylonitrile is found by SCHMIEDER and WOLF (1953) to exhibit multiple loss peaks, the crystalline melting region not being reached for this polymer before decomposition occurs. Definite peaks for frequencies of 5.2 and 4.4 cps are found at 380 and 410° K; in the region from 220 to 320° K one or more peaks also occur. The double peak in the 370 to 420° K region has been attributed to motions in the amorphous regions. A 1-1 copolymer of vinyl chloride and acrylonitrile investigated by WOLF and SCHMIEDER shows a crystalline melting peak, a large amorphous peak and a smaller secondary dispersion at lower temperatures.

DUNELL and PRICE studied the dynamic mechanical properties of viscose rayon under forced longitudinal vibrations in the 8—80 cps frequency range from 273 to 190° K finding a dispersion region around 230 to 240° K which they attributed to either motions of a single glucose unit or non-hydrogen bonded —CH_2OH groups in amorphous peaks.

SCHMIEDER and WOLF (1952) have investigated nitrocellulose plasticized with dibutyl phthalate at low frequencies (~ 0.1 to 10 cps). Unplasticized nitrocellulose exhibits a secondary damping maximum around 273° K attributed to motions in the amorphous regions. Both this maximum and the primary maximum due to crystalline melting move to lower temperatures as the amount of platicizer is increased.

Both highly crystalline and completely amorphous samples of polypropylene have been studied recently by R. A. WALL (unpublished results) over the 80 to 425° K range at audiofrequencies. The amorphous specimen exhibited a sharp primary transition at about 270° K (1200 cps) with a low broad secondary peak occurring at around 235° K (1500 cps). For the crystalline sample similar peaks at 300° K (1200 cps) and 250° K (1800 cps) were found as well as a third dispersion at 425° K (100 cps). The presence and shape of the two peaks for the amorphous material are reminiscent of those found for poly-(vinyl chloride), poly-(vinyl fluoride) and polytrifluoromonochloroethylene (see Fig. 16); since there are structural similarities between the four polymers, the secondary damping peak for all may be due to onset of a similar type of motion.

Bibliography

1. ALFREY JR., T., N. WIEDERHORN, R. STEIN and A. TOBOLSKY: Some studies of plasticized polyvinyl chloride. J. Colloid Sci. **4**, 211—227 (1949).
2. BACCAREDDA, M., and E. BUTTA: Sound velocity, Young's modulus and transition temperature in polythenes with different degree of crystallinity. J. Polymer Sci. **22**, 217—222 (1956).
3. BECKER, G.W.: Mechanische Relaxationserscheinungen in nichtweichgemachten hochpolymeren Kunststoffen. Kolloid-Z. **140**, 1—31 (1955).
4. —, u. H. OBERST: Über das dynamisch-elastische Verhalten von Polyamiden und Polyurethanen in Abhängigkeit von Frequenz, Temperatur und Wassergehalt. Kolloid-Z. **152**, 1—8 (1957).
5. BECKETT, C. W., K. S. PITZER and R. SPITZER: The thermodynamic properties and molecular structure of cyclohexane, methylcyclohexane, ethylcyclohexane and the seven dimethylcyclohexanes. J. Amer. chem. Soc. **69**, 2488—2495 (1947).
6. BLANCHETTE, J. A., and L. E. NIELSEN: Characterization of Graft Copolymers. J. Polymer Sci. **20**, 317—325 (1956).
7. BRILL, R. J.: Über das Verhalten von Polyamiden beim Erhitzen. J. prakt. Chem. **161**, 49—64 (1942).
8. BUCHDAHL, R., and L. E. NIELSEN: Multiple dispersion regions in rigid polymeric systems. J. Polymer Sci. **15**, 1—7 (1955).
9. BUTTA, E.: Sound velocity and damping in Ziegler polyethylene. J. Polymer Sci. **25**, 239—242 (1957).
10. CHARLESBY, A.: Cross-linking of polythene by pile radiation. Proc. roy. Soc. A **215**, 187—214 (1952).
11. — Effect of high-energy radiation on long-chain polymers. Nature (Lond.) **171**, 167 (1953).
12. —, and N. H. HANCOCK: The effect of cross-linking on the elastic modulus of polythene. Proc. roy. Soc. A **218**, 245—255 (1953).
13. —, and M. ROSS: The effect of cross-linking on the density and melting of polythene. Proc. roy. Soc. A **217**, 122—135 (1953).
14. CHILD JR., W. C., and J. D. FERRY: Dynamic mechanical properties of poly-n-butyl methacrylate. J. Colloid Sci. **12**, 327—341 (1957).
15. CHIPIRO, A.: Action des rayons γ sur les polymères à l'état solide. J. Chim. phys. **52**, 246—257 (1955).

16. Deeley, C. W., D. E. Kline, J. A. Sauer and A. E. Woodward: Effect of pile irradiation on the dynamic mechanical properties of polyethylene. J. Polymer Sci. **28**, 109—120 (1958).
17. —, A. E. Woodward and J. A. Sauer: Effect of irradiation on dynamic mechanical properties of 6-6 nylon. J. appl. Phys. **28**, 1124—1130 (1957).
18. Deutsch, K., E. A. W. Hoff and W. Reddish: Relation between the structure of polymers and their dynamic mechanical and electrical properties. Part I. Some alpha-substituted acrylic ester polymers. J. Polymer Sci. **13**, 565—581 (1954).
19. Dunell, B. A., and S. J. W. Price: Dispersion of mechanical properties of viscose rayon at low temperature. J. Polymer Sci. **18**, 305—306 (1955).
20. Ferry, J. D.: Mechanical properties of substances of high molecular weight. VI. Dispersion in concentrated polymer solutions and its dependence on temperature and concentration. J. Amer. chem. Soc. **72**, 3746—3752 (1950).
21. — Structure and mechanical properties of plastics. Die Physik der Hochpolymeren. 4. Band, pp. 373—419. Berlin-Göttingen-Heidelberg: Springer 1956.
22. —, W. C. Child jr., R. Zand, D. M. Stern, M. L. Williams and R. F. Landel: Dynamic mechanical properties of polyethyl methacrylate. J. Colloid Sci. **12**, 53—67 (1957).
23. —, L. D. Grandine jr. and E. R. Fitzgerald: The relaxation distribution function of polyisobutylene in the transition from rubber-like to glass-like behavior. J. appl. Phys. **24**, 911—916 (1953).
24. Fitzgerald, E. R.: Mechanical resonance dispersion in crystalline polymers at audio-frequencies. J. chem. Phys. **27**, 1180—1193 (1957).
25. —, L. D. Grandine jr., and J. D. Ferry: Dynamic mechanical properties of polyisobutylene. J. appl. Phys. **24**, 650—655 (1953).
26. Fukada, E.: On the relation between creep and vibrational loss of polymethyl methacrylate. J. Phys. Soc. Japan **6**, 254—256 (1951).
27. — The relation between dynamic elastic modulus, internal friction, creep and stress relaxation in polymethyl methacrylate. J. Phys. Soc. Japan **9**, 786—789 (1954).
28. Fuoss, R. M.: Electrical properties of solids. VI. Dipole rotation in high polymers. J. Amer. chem. Soc. **63**, 369—378 (1941).
29. Hartmann, A.: Über die Wechselwirkung zwischen Polyvinyl-Chlorid und Weichmachern. Kolloid-Z. **148**, 30—36 (1956).
30. Heijboer, J.: Molekulare Deutung sekundärer Dämpfungsmaxima-Bewegungen von Atomgruppen in Polymethacrylaten im Glaszustand. Kolloid-Z. **148**, 36—46 (1956).
31. —, P. Dekking and A. J. Staverman: The secondary maximum in the mechanical damping of polymethyl methacrylate: Influence of temperature and chemical modification. Proc. 2d Intern. Congr. Rheology (1953), pp. 123—132 London: Butterworth 1954.
32. Hellwege, K. H., R. Kaiser u. K. Kuphal: Kristallinität und mechanische Dämpfung von Polyäthylenen. Kolloid-Z. **147**, 155—156 (1956).
33. Hoff, E. A. W., D. W. Robinson and A. H. Willbourn: Relation between the structure of polymers and their dynamic mechanical and electrical properties. Part II. Glassy state mechanical dispersions in acrylic polymers. J. Polymer Sci. **18**, 161—175 (1955).
34. Iwayanagi, S.: (1) Dynamical measurements on the viscoelasticity of polymethyl acrylate. J. Sci. Res. Inst. (Tokyo) **49**, 13—22 (1955).
35. — (2) On the viscoelastic properties of linear amorphous polymers in relation to their molecular structure (polymethyl methacrylate and acrylate). J. Sci. Res. Inst. (Tokyo) **49**, 23—34 (1955).

36. Iwayanagi, S., and T. Hideshima: Low frequency coupled oscillator and its application to high polymer study. J. Phys. Soc. (Japan) **8**, 365—368 (1953).
37. — — Dynamical study on the secondary anomalous absorption region of polymethyl methacrylate. J. Phys. Soc. (Japan) **8**, 368—371 (1953).
38. Jenckel, Ernst: Zur Schwingungsdämpfung an Hochpolymeren. Kolloid-Z. **136**, 142—152 (1954).
39. —, u. H. U. Herwig: Schwingungsdämpfung und Einfriertemperaturen in Mischpolymerisaten, Polymerisatgemischen und Lösungen. Kolloid-Z. **148**, 57—66 (1956).
40. —, u. K.-H. Illers: Über die Temperaturabhängigkeit der inneren Dämpfung von weichgemachtem Polymethacrylsäuremethylester. Z. Naturforsch. **9** A, 440—450 (1954).
41a. Kabin, S. P.: On the dynamic mechanical properties of polyethylene and polytetrafluoroethylene. Zh. Tech. Fiz. USSR **26**, 2628—2632 (1956). (In Russian.)
41b. —, and G. D. Mikhailov: Mechanical and dielectric losses of polyisobutylene. Zh. tekh. Fiz. USSR **26**, 511—515 (1956). (In Russian.)
42. Karpovich, J.: Investigation of rotational isomers with ultra-sound. J. chem. Phys. **22**, 1767—1773 (1954).
43. Kawaguchi, T.: The dielectric and dynamic mechanical properties of polycaproamide. Chem. High Polymers (Japan) **13**, 283—287 (1956). (In Japanese.)
44. Kline, D. E., J. A. Sauer and A. E. Woodward: Effect of branching on dynamic mechanical properties of polyethylene. J. Polymer Sci. **22**, 455—462 (1956).
45. Krishnamurthi, M., and S. Sastry: Transition temperature of polymers at ultrasonic frequencies. Nature (Lond.) **174**, 132—133 (1954).
46a. Lawton, E. J., A. M. Bueche and J. S. Balwit: Irradiation of polymers by high-energy electrons. Nature (Lond.) **172**, 76—77 (1953).
46b. McCrum, N. G.: The low temperature transition in polytetrafluoroethylene. J. Polymer Sci. **27**, 555—558 (1958).
47. Maxwell, B.: An investigation of the dynamic mechanical properties of polymethyl methacrylate. J. Polymer Sci. **20**, 551—563 (1956).
48. Merz, E. H., L. E. Nielsen and R. Buchdahl: Influence of molecular weight on the properties of polystyrene. Ind. Engng. Chem. **43**, 3196—1401 (1951).
49a. Mikhailov, G. P.: Influence of orientation on dielectric losses of polar polymers. Zhur. Tekh. Fiz. USSR **21**, 1395—1401 (1951). (In Russian.)
49b. Mikhailov, I. G., and I. G. Solov'ev: Investigation of the mechanical properties of polyethylene and paraffin by the composite dipole method. J. Acoustics (USSR) **3**, 65—73 (1957). (In Russian.)
50. Nielsen, L. E.: Effect of crystallinity on the dynamic mechanical properties of polyethylenes. J. appl. Phys. **25**, 1209—1212 (1954).
51. Oakes, W. G., and D. W. Robinson: Dynamic electrical and mechanical properties of polyethylene over a wide temperature range. J. Polymer Sci. **14**, 505—507 (1954).
52. Pierce jr., R. H. H., E. S. Clark, J. F. Whitney and W. M. D. Bryant: Crystal structure of polytetrafluoroethylene. Paper presented before the Division of Polymer Chemistry at the 130th Meeting of the American Chemical Society, Atlantic City, N. J., Sept. 1956.
53. Sauer, J. A., N. Fuschillo, C. W. Deeley and A. E. Woodward: Segmental motion in polymers below 200° K. Proceedings of the International Symposium on Low Temperature Chemistry and Physics, Madison, Wisconsin, September 1957.

54. SAUER, J. A., and D. E. KLINE: Dynamic mechanical properties of polystyrene, polyethylene and polytetrafluoroethylene at low temperatures. J. Polymer Sci. **18**, 491—495 (1955).
55. — — Relaxation properties of high polymers. International Congress of Applied Mechanics, Brussels, Belgium. September 1956. Vol. **5**, pp. 368 — 1957.
56. SCHMIEDER, K., u. K. WOLF: Über die Temperatur- und Frequenzabhängigkeit des mechanischen Verhaltens einiger hochpolymeren Stoffe. Kolloid-Z. **127**, 65—78 (1952).
57. — — Mechanische Relaxationserscheinungen an Hochpolymeren (Beziehungen zur Struktur). Kolloid-Z. **134**, 149—184 (1953).
58a. SCHULZ, A. K.: Sur la relaxation méchanique des matières plastiques. J. Chim. phys. **53**, 933—938 (1956).
58b. SINNOT, K. M.: An apparatus for the measurement of torsion modulus and internal friction between 4.2 and 100° K. Paper presented before the High Polymer Division of the American Physical Society, March 1958, Chicago, Illinois. To be published.
59. SUBRAHMANGAM, S. V.: Temperature dependence of ultrasonic velocity in plastics. J. chem. Phys. **22**, 1562—1653 (1954).
60. TILL JR., P. H.: The growth of single crystals of linear polyethylene. J. Polymer Sci. **24**, 301—306 (1957).
61. THOMAS, D. A., and D. W. ROBINSON: Some dynamic mechanical properties of polyisobutylene over a wide temperature range. Brit. J. appl. Phys. **6**, 41—43 (1955).
62. THOMPSON, A. B., and D. W. WOODS: The transitions of polyethylene terephthalate. Trans Faraday. Soc. **52**, 1383—1397 (1956).
63. THURN, H.: Dielektrizitätskonstante und mechanische Verluste bei Hochpolymeren. Z. angew. Phys. **7**, 44—47 (1955).
64. —, u. K. WOLF: Vergleichende dielektrische und Ultraschall-Messungen bei $2 \cdot 10^6$ hz an Polyvinylestern, Polyacrylestern und Polyvinyläthern. Kolloid-Z. **148**, 16—30 (1956).
65. TRELOAR, L. R. G.: The structure and mechanical properties of rubberlike materials. Die Physik der Hochpolymeren. 4. Band, pp. 294—372. Berlin-Göttingen-Heidelberg: Springer 1956.
66. WADA, Y., and K. YAMAMOTO: Temperature dependence of velocity and attentuation of ultrasonic waves in high polymers. J. Phys. Soc. Japan **11**, 887—892 (1956).
67. WALTER, A. T.: Elastic properties of polyvinyl chloride gels. J. Polymer Sci. **13**, 207—227 (1954).
68. WOLF, K., and K. SCHMIEDER: Mechanical-thermal transition ranges in partially crystalline macromolecules and their relationship to structure. An extract from the records of the "International Symposium on Macromolecular Chemistry"; a supplement to Ric. sci. **25** (1955) (published in USA by Interscience Inc.).
69. WOODS, D. W.: Effect of crystallization on the glass-rubber transition in polyethylene terphthalate filaments. Nature (Lond.) **174**, 753—754 (1954).
70. WOODWARD, A. E., C. W. DEELEY, D. E. KLINE and J. A. SAUER: Effect of thermal history on the dynamic modulus at 20° C of irradiated polyethylene. J. Polymer Sci. **26**, 383—386 (1957).
71. —, J. A. SAUER, C. W. DEELEY and D. E. KLINE: The dynamic mechanical behavior of some nylons. J. Colloid Sci. **12**, 363—377 (1957).
72. WÜRSTLIN, F.: Dielektrische Messung der Weichmachung von Polyvinylchlorid. Kolloid-Z. **105**, 9—16 (1943).
73. YAMAMOTO, K., and Y. WADA: Investigation of dynamic mechanical properties of glassy polymers by composite oscillator method. J. Phys. Soc. Japan **12**, 374—378 (1957).

Zeitfracht Medien GmbH
Ferdinand-Jühlke-Straße 7
99095 Erfurt, Deutschland
produktsicherheit@kolibri360.de